ORDER NUMBER EA-FAA-T-8080-12C

AVIATION MECHANIC AIRFRAME QUESTION BOOK

AF774529

1990

U.S. DEPARTMENT OF TRANSPORTATION

FEDERAL AVIATION ADMINISTRATION

Aviation Standards National Field Office

International Standard Book Number 0-89100-349-5
For sale by: IAP, Inc.
Mail to: P.O. Box 10000, Casper, WY 82602-1000
Ship to: 7383 6WN Road, Casper, WY 82604-1835
(800) 443-9250 • (307) 266-3838 • FAX: 307-472-5106

IAP, Inc.
7383 6WN Road, Casper, WY 82604-1835

Printed in the USA

CONTENTS

PREFACE

This question book has been developed by the Federal Aviation Administration (FAA) for use by applicants who are preparing for the Aviation Mechanic Airframe written test and also for use by FAA testing centers and FAA designated written test examiners in administering the test. It is issued as FAA-T-8080-12C, Aviation Mechanic Airframe Question Book, and is available to the public from:

Superintendent of Documents
U.S. Government Printing Office
Washington, DC 20402

or from U.S. Government Printing Office bookstores located in major cities throughout the United States.

The questions included in this publication are predicated on regulations, principles, and practices that were valid at the time of publication. The question selection sheets prepared for use with this question book are security items and are revised at frequent intervals.

The FAA does not supply the correct answers to questions included in this book. Students should determine the answers by research and study, by working with instructors, or by attending ground schools. The FAA is NOT responsible for either the content of commercial reprints of this book or the accuracy of the answers they may list.

Comments regarding this publication should be directed to:

U.S. Department of Transportation
Federal Aviation Administration
Aviation Standards National Field Office
Maintenance Support Branch
Testing Standards Development Section, AVN-142
P.O. Box 25082
Oklahoma City, OK 73125

MAXIMUM TIME ALLOWED FOR TEST: 4 HOURS

MATERIALS

Materials to be used with this question book when used for airman testing:

1. Airman Written Test Application which includes the answer sheet.
2. Question Selection Sheet which identifies the questions to be answered.
3. Plastic overlay sheet which can be placed over electrical drawings, graphs, and charts for making pencil marks or for tracing schematics for analysis purposes. If you are not provided the overlay, request one from the test monitor.

GENERAL INSTRUCTIONS

1. Read the instructions on page 1 of AC Form 8080-3, Airman Written Test Application, and complete the form on page 4.
2. The questions in this question book are numbered consecutively beginning with No. 6001. Refer to the question selection sheet to determine which questions to answer.
3. For each item on the answer sheet, find the appropriate question in the book.
4. Mark your answer in the space provided for that item on the answer sheet.
5. Read each question carefully and avoid hasty assumptions. Do not answer until you understand the question. Do not spend too much time on any one question. Answer all of the questions that you readily know and then reconsider those you find difficult.

   If a regulation is changed after this question book is printed, you will receive credit until the affected questions are revised.

THE MINIMUM PASSING GRADE IS 70.

WARNING

§65.18 Written tests: Cheating or other unauthorized conduct.

(a) Except as authorized by the Administrator, no person may—

(1) Copy, or intentionally remove, a written test under this part;
(2) Give to another, or receive from another, any part or copy of that test;
(3) Give help on that test to, or receive help on that test from, any person during the period that test is being given;
(4) Take any part of that test in behalf of another person;
(5) Use any material or aid during the period that test is being given; or
(6) Intentionally cause, assist, or participate in any act prohibited by this paragraph.

(b) No person who commits an act prohibited by paragraph (a) of this section is eligible for any airman or ground instructor certificate or rating under this chapter for a period of 1 year after the date of that act. In addition, the commission of that act is a basis for suspending or revoking any airman or ground instructor certificate or rating held by that person.

AVIATION MECHANIC AIRFRAME
QUESTION BOOK

INTRODUCTION

The requirements for a mechanic certificate and ratings, and the privileges, limitations, and general operating rules for certificated mechanics are prescribed in Federal Aviation Regulation (FAR) Part 65, Certification: Airmen Other Than Flight Crewmembers. Any person who applies and meets the requirements is entitled to a mechanic certificate.

At FAA testing centers, or at an FAA designated written test examiner's facility, the applicant is issued a "clean copy" of this question book, an appropriate 100-item question selection sheet which indicates the specific questions to be answered, and AC Form 8080-3, Airman Written Test Application, which contains the answer sheet. The question book contains all the supplementary material required to answer the questions. Supplementary material, such as an illustration, will normally be found within one page of the question with which it is associated.

THE WRITTEN TEST

Questions and Scoring

The test questions are of the multiple-choice type. Answers to questions listed on the question selection sheet should be marked on the answer sheet of AC Form 8080-3. Directions should be read carefully before beginning the test. Incomplete or erroneous personal information entered on this form delays the scoring process.

The answer sheet is sent to the Mike Monroney Aeronautical Center in Oklahoma City where it is scored by a computer. Shortly thereafter AC Form 8080-2, Airman Written Test Report, is sent to the applicant listing the score and test questions missed.

The applicant must present this report for an oral and practical test, or for retesting in the event of written test failure or expiration.

Taking the Test

The test may be taken at FAA testing centers, FAA written test examiners' facilities, or other designated places. After completing the test, the applicant must surrender the issued question book, question selection sheet, answer sheet, and any papers used for computations or notations to the monitor before leaving the test room.

When taking the test, the applicant should keep the following points in mind:

1. Answer each question in accordance with the latest regulations and procedures.
2. Read each question carefully before looking at the possible answers. You should clearly understand the problem before attempting to solve it.
3. After formulating an answer, determine which of the alternatives most nearly corresponds with that answer. The answer chosen should completely resolve the problem.
4. From the answers given, it may appear that there is more than one possible answer; however, there is only one answer that is correct and complete. The other answers are either incomplete or are derived from popular misconceptions.
5. If a certain question is difficult for you, it is best to proceed to other questions. After the less difficult questions have been answered, return to those which gave you difficulty. Be sure to indicate on the question selection sheet the questions to which you wish to return.
6. When solving a computer problem, select the answer nearest your solution. The problem has been checked with various types of computers; therefore, if you have solved it correctly, your answer will be closer to the correct answer than to any of the other choices.
7. To aid in scoring, enter personal data in the appropriate spaces on the test answer sheet in a complete and legible manner. Enter the test number printed on the question selection sheet.

Retesting—FAR 65.19

Applicants who receive a failing grade may apply for retesting by presenting their AC Form 8080-2, Airman Written Test Report—

(1) after 30 days from the date the applicant failed the test; or

(2) the applicant may apply for retesting before the 30 days have expired upon presenting a signed statement from an airman holding the certificate and rating sought by the applicant, certifying that the airman has given the applicant additional instruction in each of the subjects failed and that the airman considers the applicant ready for retesting.

NOTE: Blank spaces appear on several pages of this publication. This permits referenced figures and tables to appear as close as possible to their related questions.

AVIATION MECHANIC AIRFRAME
QUESTION BOOK

SECTION 1 — AIRFRAME STRUCTURES

6001. Laminated wood spars may be substituted for solid rectangular wood spars

1—only in certain instances where the primary load is shared by one or more other original structural member.
2—only if the strength deficiency is allowed for by an increase in dimension.
3—if the same quality wood is used in both.
4—only upon specific approval by the manufacturer or the FAA.

6002. The strength of a glue joint must be equal to

1—75 percent of the wood strength.
2—the strength of the wood.
3—twice the strength of the wood in tension.
4—50 percent of the wood strength.

6003. Elongated boltholes in a wood spar may be repaired by

1—splicing in a new section of spar and drilling new holes.
2—attaching reinforcement plates of one-fourth the spar width.
3—reaming the holes and inserting bushings whose wall thickness is one-fourth the original bolt diameter and whose length is one thirty-second less than the spar width.
4—reaming and inserting the next standard size larger bolt.

6004. The strength of a well-designed and properly prepared wood splice joint is provided by the

1—bearing surface of the wood fibers.
2—doubler.
3—glue.
4—reinforcement plates.

6005. Where is information found concerning acceptable species substitutions for wood materials used in aircraft repair?

1—Aircraft Specifications or Type Certificate Data Sheets.
2—Technical Standard Orders.
3—AC 43.13-1A.
4—AC 65-19C.

6006. In cases of elongated boltholes in a wood spar or cracks in the vicinity of boltholes,

1—it is permissible to plug the hole with hardwood and redrill.
2—a plug should be inserted in the hole and a reinforcement plate added.
3—the spar may be reinforced by using birch reinforcing plates.
4—a new section of spar should be spliced in or the spar replaced entirely.

6007. A faint line running across the grain of a wood spar generally indicates

1—compression failure.
2—fungus growth.
3—shear failure.
4—decay.

6008. Which statement about wood decay is correct?

1—Decay that occurs before the wood is seasoned does not affect the strength of the finished piece.
2—Decay that does not cause the wood to discolor or stain is not usually harmful to the strength of the wood.
3—A limited amount of certain kinds of decay is acceptable in aircraft woods since decay affects the binding between the fibers and not the fibers themselves.
4—Decay, unlike some other wood defects, is not acceptable in any form or amount.

6009. Which of the following conditions will determine acceptance of wood with mineral streaks?

1—Careful inspection fails to reveal any decay.
2—They do not cause grain divergence.
3—They produce only a small effect on grain direction.
4—Local irregularities do not exceed limitations specified for spiral and diagonal grain.

6010. The I-beam wooden spar is routed to

1—improve shape.
2—increase strength.
3—remove imperfections.
4—reduce weight.

6011. Pin knot clusters are permitted in wood structure provided

1— they produce a small effect on grain direction.
2— they are located 24 inches apart.
3— they have no mineral streaks.
4— no pitch pockets are near.

6012. The cantilever wing uses

1— external wood bracing.
2— ribs as the principal structural members.
3— no external bracing.
4— the skin to carry most of the load to the wing butt.

6013. Laminated wood is often used in the construction of highly stressed aircraft components. This wood can be identified by its

1— increased resistance to deflection.
2— parallel grain construction.
3— similarity to standard plywood construction.
4— alternate perpendicular ply construction.

6014. When patching a plywood skin, abrupt changes in cross-sectional areas which will develop dangerous stress concentration should be avoided by using

1— circular or elliptical patches.
2— rectangular patches.
3— square patches.
4— doublers with any desired shaped patches.

6015. When making an aircraft wood repair, special care should be exercised when determining the proper grain direction because the

1— shear strength of wood is many times higher perpendicular to the grain than parallel to the grain.
2— low tensile strength parallel to the grain should be used to the fullest advantage.
3— tensile strength of wood is many times higher perpendicular to the grain than parallel to the grain.
4— high shearing strength parallel to the grain should be used to the fullest advantage.

6016. Which type of patch is preferred for repairing stressed plywood skin?

1— Surface patch.
2— Splayed patch.
3— Plug patch.
4— Scarf patch.

6017. Compression failures in wood structures are characterized by buckling of the fibers that appear as streaks on the surface

1— at right angles to the growth rings.
2— only in the heartwood.
3— only in the springwood or summerwood.
4— at right angles to the grain.

6018. Three forms of decay in wood structures are

1— mineral streaks, compression wood, and dote.
2— compression wood, dote, and redheart.
3— dote, redheart, and purpleheart.
4— springwood, summerwood, and dote.

6019. The minimum allowable tensile strength of new Grade A (TSO-C15) Aircraft cotton fabric in pounds per inch warp and fill is

1— 100.
2— 65.
3— 46.
4— 80.

6020. Fabric coverings are laced to the wing structure for the purpose of maintaining the contour of the airfoil and to prevent lifting and whipping of the fabric in flight. The lacings should be secured by the use of seine knots

1— positioned in the middle of, or along the edge of, the bottom capstrip.
2— tied on the outside, and then pulled into the wing structure.
3— which should never extend above the adjacent wing surface.
4— positioned in the middle of the top capstrip for aircraft with a never-exceed speed of 150 MPH or less.

6021. When and how is finishing tape applied on a fabric-covered aircraft?

1— Sewed or laced on before dope is applied.
2— Doped on immediately prior to the finish coat.
3— Doped on after the first or second coat of dope.
4— Doped on at the same time the last coat of dope or lacquer is applied.

6022. The determining factor(s) for the selection of the correct weight of textile fabric to be used in covering any type of aircraft is the

1— maximum wing loading.
2— speed of the aircraft.
3— weight class of the aircraft.
4— speed of the aircraft and the maximum wing loading.

6023. How many fabric thicknesses will be found in a French-fell seam?

1— One.
2— Two.
3— Three.
4— Four.

6024. Prior to application of surface tape to the trailing edge of a wing or control surface, the tape should be notched at intervals to

1— make it easier to apply to severe curvatures.
2— increase the length of the tape's edges for better doping.
3— prevent the entire tape from loosening in the event the tape begins to separate from the surface.
4— prevent raveling of the tape.

6025. Moisture, mildew, chemicals, and acids have no effect on

1— glass fabric.
2— cotton fabric.
3— linen fabric.
4— dacron fabric.

6026. Surface or finish tape is installed on fabric-covered aircraft for what purpose?

1— Prevent loosening of the entire fabric covering.
2— Attach the fabric to the structure.
3— Prevent wear over lacings, seams, corners, and edges.
4— Prevent warping of the surface during drying period of the dope.

6027. The best method of repair for a fabric-covered surface which has an L-shaped tear, each leg of which is approximately 14 inches long, is to

1— re-cover the entire bay in which the tear is located.
2— cut out the tear, sew in a new section, and then dope the new section to conform with the remainder of the surface.
3— sew from the end of each leg to the center of the tear with a baseball stitch and then dope on a patch.
4— sew with a baseball stitch from the center of the tear out toward the extremity of each leg and then dope on a patch.

6028. The recommended procedure for making an unsewed (doped-on) fabric patch is to

1— cover the edges of all patches by applying pinked-edge surface tape with the second coat of clear dope.
2— dope-on the patch by means of initial coat of clear dope thinned 50 percent to obtain sufficient softening of the dope on the underlying fabric.
3— clean and remove all aluminum and color coats from the surface to which the patch is to be doped.
4— clean the area thoroughly and remove all color coats, aluminum, and clear dope from the surface to be covered by the patch, using acetone only.

6029. A fabric-covered surface has a damaged area that measures 19 inches in one direction. How should the damaged area be repaired?

1— Doped-on patch.
2— Dope-in a new panel.
3— Sewed-in patch.
4— Sewed-in patch with doped-on surface patch.

6030. When damage to a fabric-covered surface is such that it will permit a sewn-in repair patch, what type of stitch should be used?

1— Half-hitch.
2— Modified seine.
3— Baseball.
4— Double loop.

6031. Whenever it is necessary to install an unsewed patch, which of the following is a correct statement?

1— A repair should return the original strength and tautness to the fabric.
2— Remove old dope and paint by wire brushing down to base coat.
3— A repair should return 70 percent of the original strength and tautness to the fabric.
4— Install the patch over the color coats.

6032. (1) Machine-sewn seams in aircraft covering fabrics may be of the folded-fell or French-fell types.

(2) A plain lapped seam is never permissible.

Regarding the above statements,

1— both No. 1 and No. 2 are true.
2— neither No. 1 nor No. 2 is true.
3— only No. 1 is true.
4— only No. 2 is true.

6033. Thread for handsewing and lacing cord should be waxed before using. The wax should not exceed what percent of the weight of the finished cord?

1—20.
2—25.
3—30.
4—50.

6034. When dope proofing the parts of the aircraft structure that come in contact with doped fabric, how many of the following provide an acceptable protective coating?

A. Aluminum foil.
B. Cellulose tape.
C. Zinc chromate primer.
D. Resin impregnated cloth tape.

1—One.
2—Two.
3—Three.
4—Four.

6035. The letters and numbers of identification markings must be of a contrasting color and must be uniform in shape and size. The width of the characters must be

1—one-half their height.
2—two-thirds their height.
3—equal to their height.
4—five times the width of the stroke.

6036. Which statement is correct concerning aircraft registration numbers?

1—All registered aircraft operating in the United States must display a registration number.
2—Restricted category aircraft operating in the United States are not required to display a registration number.
3—All aircraft located in the United States must display a registration number.
4—Only normal and utility category aircraft must display a registration number.

6037. It is inadvisable to dope fabric on humid days because this increases the tendency of the dope to

1—pebble.
2—pull and rope.
3—run.
4—blush.

6038. When finishing a fabric surface, apply

1—a minimum of eight coats.
2—at least the number of coats necessary to result in a taut and well-filled finish job.
3—the first three coats of dope by brush and the remaining five coats by spray gun.
4—three coats of clear, two coats of aluminum, and four coats of pigmented dope.

6039. A practical method of determining whether a fabric surface has been finished with nitrate (cellulose nitrate) or butyrate (cellulose acetate butyrate) dope is to

1—press the thumb firmly against an unsupported section of the fabric and note the time required for the fabric to recover its original shape.
2—rub a small section with lacquer thinner. Butyrate finish will not be affected, but nitrate finish will be softened.
3—remove a 1-inch-wide strip of fabric and conduct a tensile test.
4—observe the reverse side of the fabric for evidence of blue dye (butyrate) or red dye (nitrate) that has penetrated as a result of brushing on the first two coats of dope.

6040. The chief purpose of pigment (or aluminum) in dope is to

1—make it easier to apply.
2—make a more pleasing appearance.
3—give a covering that has a minimum of air resistance.
4—exclude sunlight from the fabric.

6041. A correct use for acetone is to

1—thin shellac.
2—thin zinc chromate primer.
3—remove grease from fabric.
4—thin dope.

6042. Whenever operations involving sanding, rubbing, or wiping of detached fabric-covered wings are to be performed, the surface should always be

1—coated with water to prevent overheating of the finish during the finishing process.
2—coated with water to prevent scratching of the finish during the finishing process.
3—located in a room in which the relative humidity is not more than 20 percent. By closely controlling the relative humidity, the material used for sanding, rubbing, or wiping will remain dry during the entire operation.
4—grounded to prevent accumulation of static charges during finishing process.

6043. Why is the first coat of dope brushed onto aircraft fabric and not sprayed on?

1—To permit thorough penetration of the fabric.
2—To provide slow shrinkage of the fabric and thereby ensure a taut surface.
3—To provide a thick base which will prevent peeling of subsequent applications.
4—To build up a sufficiently thick surface layer and thereby eliminate the need for excessive doping.

6044. Fungicidal dopes are used in aircraft finishing as the

1—first coat to prevent fabric rotting and are applied extremely thin to saturate the fabric.
2—first full-bodied brushed-on coat to prevent fungus damage.
3—finish clear coat to provide a glossy appearance and to prevent deterioration of the fabric.
4—final full-bodied brushed-on coat to reduce blushing.

6045. Before spraying a paint finish on cleaned aluminum, never

1—rinse off the surface with water.
2—touch the surface with bare hands.
3—blow off the surface with air.
4—suspend the surface in a dry room.

6046. What is the usual cause of runs and sags in aircraft finishes?

1—Too much material applied in one coat.
2—Material drying too fast.
3—Material is being applied too fast.
4—High atmospheric humidity.

6047. Which defect in aircraft finishes may be caused by moisture in the air supply line, adverse humidity, drafts, or sudden changes in temperature?

1—Spray dust.
2—Spray mottle.
3—Blushing.
4—Sags and runs.

6048. Which statement is true regarding paint system compatibility?

1—Old-type zinc chromate primer may not be used directly for touchup of bare metal surfaces.
2—Acrylic nitrocellulose lacquers may be used over old nitrocellulose finishes.
3—Old wash primer coats may be overcoated directly with epoxy finishes.
4—Modified zinc chromate primer will adhere satisfactorily to bare metal.

6049. A well-designed rivet joint will subject the rivets to

1—compressive loads.
2—shear loads.
3—tension loads.
4—bending loads.

6050. If a Hi-Shear rivet is underdriven during installation, the

1—stud will not expand and fill the drilled hole properly.
2—collar will be incompletely swaged into the groove.
3—shear strength of the rivet will be reduced.
4—stem will loosen, resulting in a hollow rivet shank.

6051. Alloy 2117 rivets are heat treated

1—by the manufacturer and do not require heat treatment before being driven.
2—by the manufacturer but require reheat treatment before being driven.
3—to a temperature of 910 to 930 °F and quenched in cold water.
4—to a temperature of 930 to 950 °F and quenched in cold water.

6052. The general rule for finding the proper rivet diameter is

1—three times the thickness of the thinnest sheet.
2—three times the thickness of the materials to be joined.
3—two times the rivet length.
4—three times the thickness of the thickest sheet.

6053. The bucked head of a rivet should be

1—one and one-half D by one-half D.
2—one-half D by one D.
3—one and one-half D by two D.
4—two and one-half D by one D.

6054. Pin (Hi-Shear) rivets have the same shear strength as

1—solid shank AN470DD (icebox) rivets of the same diameter.
2—bolts of equal diameter.
3—hollow-core rivets of the same diameter.
4—self-plugging friction lock rivets of equal diameter.

6055. The markings on the head of a Dzus fastener identify the

1—body diameter, type of head, and length of the fastener.
2—body type, head diameter, and type of material.
3—manufacturer and type of material.
4—manufacturer and distributor.

6056. The Dzus turnlock fastener consists of a stud, grommet, and receptacle. The stud length is measured in

1—hundredths of an inch.
2—tenths of an inch.
3—thousandths of an inch.
4—sixteenths of an inch.

6057. The Dzus turnlock fastener consists of a stud, grommet, and receptacle. The stud diameter is measured in

1—tenths of an inch.
2—hundredths of an inch.
3—sixteenths of an inch.
4—thousandths of an inch.

6058. Self-plugging Cherry rivet shank diameters are measured in increments of

1—1/16 inch.
2—3/32 inch.
3—1/8 inch.
4—1/32 inch.

6059. Bulbed Cherrylock rivet grip length is measured in increments of

1—1/32 inch.
2—1/16 inch.
3—3/32 inch.
4—1/8 inch.

6060. Sandwich panels made of metal honeycomb construction are used in high-speed aircraft because this type construction

1—is lighter than single sheet skin of the same strength and is more corrosion resistant.
2—may be repaired by gluing replacement skin to the inner core material with thermoplastic resin.
3—requires only self-tapping screws for the attachment of skin when repairing damaged areas.
4—has a high strength-to-weight ratio and greater stiffness than a single sheet.

6061. The simplest test in checking for a delaminated area on a honeycomb structure would be

1—the metallic ring test.
2—x-ray.
3—ultrasonic.
4—zyglo.

6062. When removing damage from one face sheet of a laminated honeycomb panel, it is important to avoid damaging the opposite face sheet. The most effective tool for removing a damaged area 1-inch diameter or less is a

1—router and template.
2—hole saw of the proper size to remove all the damage.
3—rotary file.
4—pair of aviation snips for a left-hand cut.

6063. When balsa wood is used to replace a damaged honeycomb core, the plug should be cut so that

1—the grain is parallel to the skin.
2—it will be slightly compressed when the repair is completed.
3—it is about 1/8 inch undersize to allow sufficient bonding material to be applied.
4—the grain is perpendicular to the skin.

6064. When repairing puncture-type damage of a metal faced laminated honeycomb panel, the edges of the doubler should be tapered

1—two times the thickness of the metal.
2—five times the thickness of the metal.
3—100 times the thickness of the metal.
4—to whatever is desired for appearance.

6065. What is the most effective method for determining if the proper ratio of resin-to-catalyst was used when making epoxy repairs?

1—Make a pull test on the overlapping material.
2—Test with a hardness tester in each square inch of new material.
3—Mix enough epoxy to have a sample for testing after curing.
4—Mixing ratios are not critical, therefore, no evaluation of the materials used is necessary.

6066. How can stress concentrations be eliminated when making scarf repairs in composite structures?

1— Use a core filter of higher density than the original material.
2— Mix the resin catalyst in such proportions that the cured resin will be somewhat flexible.
3— Avoid abrupt changes in cross-sectional areas.
4— Use square- or rectangular-shaped scarf patches.

6067. What precaution, if any, should be taken to prevent corrosion inside a repaired metal honeycomb structure?

1— Prime the repair with a corrosion inhibitor and seal from the atmosphere.
2— Paint the outside area with several coats of exterior paint.
3— Coat mating surfaces and all fasteners with thin oil.
4— None. Honeycomb is usually made from a man-made or fibrous material which is not susceptible to corrosion.

6068. One method of inspecting a laminated fiberglass structure that has been subjected to damage is to

1— strip the damaged area of all paint and shine a strong light through the structure.
2— use dye-penetrant inspection procedures, exposing the entire damaged area to the penetrant solution.
3— use an eddy current probe on both sides of the damaged area.
4— use magnaflux inspection procedures for all areas where damage is suspected.

6069. When inspecting a fiberglass panel using the "tapping" test procedure, a dull thud from the tapping tool indicates

1— a solid inner core of the laminated panel.
2— separation of the laminates used in the lay-up.
3— that the fiberglass resin is properly cured.
4— that the panel is probably serviceable.

6070. How many of the following are benefits of using microballoons when making repairs to laminated honeycomb panels?

A. Less filler shrinkage.
B. Greater flexibility.
C. Less density.
D. Lower stress concentrations.

1— One.
2— Two.
3— Three.
4— Four.

6071. Damage evaluation charts for honeycomb structures of a large aircraft can be found in the applicable section of its

1— overhaul manual.
2— maintenance manual.
3— structural repair manual.
4— Type Certificate Data Sheet.

6072. Which is an identifying characteristic of cellulose acetate plastics?

1— Burns with a steady, clear flame.
2— Rub and blow acetone; it will turn blue.
3— When heated or burned, has a very pleasant odor.
4— Zinc chloride will turn the plastic milky.

6073. When inspecting a honeycomb structure using the metallic ring test, a clear metallic sound indicates

1— core damage.
2— the facing and core are sound.
3— facing damage.
4— facing and core separation.

6074. Which is an identifying characteristic of acrylic plastics?

1— When heated or burned, has a very repugnant odor.
2— Zinc chloride will have no effect.
3— Has a yellowish tint when viewed from the edge.
4— Acetone will soften plastic, but will not change its color.

6075. Superficial scars, scratches, surface abrasion, or rain erosion on fiberglass laminates can generally be repaired by applying

1—a piece of resin-impregnated glass fabric facing.
2—a surface patch by means of epoxy resin cured for 1 hour with an infrared heat lamp.
3—one or more coats of suitable resin (room-temperature catalyzed) to the surface.
4—a sheet of cellophane over the abraded surface and one or more coats of resin cured with infrared heat lamps.

6076. One of the main advantages of honeycomb structure is that it

1—is better able to withstand sonic vibration.
2—is relatively fireproof.
3—does not require inspections.
4—is very strong because it is heavy.

6077. A potted compound repair on honeycomb can usually be made on damages less than

1—4 inches in diameter.
2—3 inches in diameter.
3—2 inches in diameter.
4—1 inch in diameter.

6078. Honeycomb core material used to replace damaged aluminum cores of honeycomb bonded structures is usually

1—titanium.
2—fiberglass.
3—stainless steel.
4—magnesium.

6079. In a transition area bonded honeycomb panel repair, the circular shape repair should not exceed

1—2 inches in diameter.
2—twice the bay area thickness.
3—five times the flange area thickness.
4—5 inches in any direction.

6080. Exposure of honeycomb structures to sonic vibrations usually causes

1—delaminations.
2—radiographic corrosion.
3—aerodynamic smoothness.
4—punctures.

6081. A typical repair used to retain 100 percent strength to a damaged area 1 inch in diameter of bonded honeycomb is

1—a step-type metal facing patch.
2—an overlap-type metal facing patch.
3—a scarf-type facing patch.
4—an overlap-type resin-glass facing patch.

6082. What type inspection procedure should be used to check for water in the honeycomb of a composite structure?

1—Shake the component and listen for water movement.
2—Eddy current.
3—Magnetic particle.
4—Radiography.

6083. Polyester resin is cured by which of the following processes?

1—Applying external heat.
2—Air cure.
3—Adding a catalyst which joins the molecules.
4—A combination of air drying and pressure application.

6084. When repairing large, flat surfaces with polyester resins, warping of the surface is likely to occur. One method of reducing the amount of warpage is to

1—add an extra amount of catalyst to the resin.
2—use short strips of fiberglass in the bonded repair.
3—make the repair using heavy layers of fiberglass cloth.
4—use less catalyst than normal so the repair will be more flexible.

6085. When making repairs to fiberglass structures, cleaning of the area to be repaired is essential for a good bond. The final cleaning should be made using

1—lacquer thinner.
2—methyl ethyl ketone (MEK).
3—soap, water, and a scrub brush.
4—a thixotropic agent.

6086. What is the purpose of a thixotropic agent used in plastic resins?

1—Prolongs the curing process.
2—Reduces the moisture content of resin.
3—Adds flexibility to the resin.
4—Thickens the resin to a paste-like consistency.

6087. Fiberglass laminate damage not exceeding the first layer or ply can be repaired by

1—filling with a putty consisting of a compatible resin and clean, short glass fibers.
2—placing a sheet metal plate over the damaged area and securing with self-tapping screws.
3—sanding the damaged area until aerodynamic smoothness is obtained.
4—trimming the rough edges and sealing with paint.

6088. Fiberglass damage that extends completely through a laminated sandwich structure

1—may be repaired.
2—may not be repaired.
3—must be filled with resin to eliminate dangerous stress concentrations.
4—may be filled with putty which is compatible with resin.

6089. Fiberglass laminate damage that extends completely through one facing and into the core

1—cannot be repaired.
2—requires the replacement of the damaged core and facing.
3—can be repaired by using a typical metal facing patch.
4—requires a plywood reinforcing plate installed with self-tapping screws over the damaged area.

6090. Which of the following is a characteristic of composites?

1—High-thermal expansion.
2—Low resistance to fatigue failure.
3—Low-thermal expansion.
4—Low resistance to corrosion.

6091. If an aircraft's transparent plastic enclosures exhibit fine cracks which may extend in a network over or under the surface or through the plastic, the plastic is said to be

1—creeping.
2—fatiguing.
3—stretching.
4—crazing.

6092. When installing transparent plastic enclosures which are retained by bolts extending through the plastic material and elastic stop nuts, the stop nuts should be

1—tightened to a firm fit plus one full turn.
2—tightened sufficiently to make a waterproof seal.
3—tightened to a firm fit then backed off one full turn.
4—tightened to a firm fit.

6093. If a new safety belt is to be installed in an aircraft, the belt must conform to the strength requirements in which document?

1—STC 1282.
2—FAR Part 65.
3—FAR Part 39.
4—TSO C22.

6094. Which is considered good practice concerning the installation of acrylic plastics?

1—When nuts and bolts are used, the plastic should be installed hot and tightened to a firm fit before the plastic cools.
2—When rivets are used, adequate spacer or other satisfactory means to prevent excessive tightening of the frame to the plastic should be provided.
3—When rivets or nuts and bolts are used, slotted holes are not recommended.
4—When nuts and bolts are used, the plastic should be installed cold and each nut tightened to a firm fit before the plastic warms up.

6095. The coefficient of expansion of most plastic enclosure materials is

1—greater than both steel and aluminum.
2—greater than steel but less than aluminum.
3—less than either steel or aluminum.
4—approximately the same as aluminum.

6096. If no scratches are visible after transparent plastic enclosure materials have been cleaned, their surfaces should be

1—buffed with a clean cloth dipped in a mixture of turpentine and chalk.
2—polished with an automobile cleanser applied with a damp cloth.
3—buffed with a clean, soft, dry cloth.
4—covered with a thin coat of wax.

6097. Cabin upholstery materials installed in current standard category airplanes must

1— be fireproof.
2— be at least flame resistant.
3— be approved in accordance with FAR Part 61.
4— meet the requirements prescribed in FAR Part 43.

6098. What is the most common method of cementing transparent plastics?

1— The heat method.
2— The soak method.
3— The splice method.
4— The bevel method.

6099. When holes are drilled completely through Plexiglas, a

1— standard twist drill may be used.
2— combination drill should be used.
3— specially modified twist drill should be used.
4— wood drill may be used.

6100. What is the purpose of a gusset or gusset plate used in the construction and repair of aircraft structures?

1— To hold structural members in position temporarily until the permanent attachment has been completed.
2— To provide access for inspection of structural attachments.
3— To join and reinforce intersecting structural members.
4— To provide a method of adjusting the tension or location of structural components.

6101. Select the alternative which best describes the function of the flute section of a twist drill.

1— Allows lubrication to reach the drill body.
2— Straightens the drilling chips to avoid clogging the hole.
3— Forms the cutting edges of the drill point.
4— Maintains the proper cooling level of the drill.

6102. How many MS20470 AD-4-6 rivets will be required to attach a 10" x 5" plate, using a single row of rivets, minimum edge distance, and 4D spacing?

1— 56.
2— 52.
3— 60.
4— 62.

6103. Longitudinal (fore and aft) structural members of a semi-monocoque fuselage are called

1— spars and ribs.
2— longerons and stringers.
3— bulkheads and rings.
4— ribs and formers.

6104. Shallow scratches in sheet metal may be repaired by

1— burnishing.
2— buffing.
3— stop drilling.
4— patching.

6105. What should be the included angle of a twist drill for soft metals?

1— 118°.
2— 90°.
3— 65°.
4— 45°.

6106. When comparing the machining techniques for stainless steel sheet material to those for aluminum alloy sheet, it is normally considered good practice to drill the stainless steel at a

1— higher speed with less pressure applied to the drill.
2— lower speed with more pressure applied to the drill.
3— higher speed with more pressure applied to the drill.
4— lower speed with less pressure applied to the drill.

6107. A single-lap sheet splice is to be used to repair a section of damaged aluminum skin. If a double row of 1/8-inch rivets is used, the minimum allowable overlap will be

1— 1-1/4 inches.
2— 1/2 inch.
3— 3/4 inch.
4— 1 inch.

6108. Which statement is true regarding the inspection of a stressed skin metal wing assembly known to have been critically loaded?

1— If rivets show no visible distortion, further investigation is unnecessary.
2— If bearing failure has occurred, the rivet shanks will be joggled.
3— If genuine rivet tipping has occurred, groups of consecutive rivet heads will be tipped in the same direction.
4— If shear failure has occurred, the sheet will be distorted from the rivet hole out to the edge of the sheet.

6109. What is the minimum edge distance for aircraft rivets?

1— Two times the diameter of the rivet shank.
2— Two times the diameter of the rivet head.
3— Three times the diameter of the rivet shank.
4— One-half the thickness of the material being riveted.

6110. When drilling stainless steel, the drill used should have an included angle of

1— 90° and turn at a high speed.
2— 90° and turn at a low speed.
3— 118° and turn at a high speed.
4— 140° and turn at a low speed.

6111. What is the minimum spacing for a single row of aircraft rivets?

1— Two times the diameter of the rivet shank.
2— Three times the length of the rivet shank.
3— One-half the thickness of the material being riveted.
4— Three times the diameter of the rivet shank.

6112. (Refer to figure 1.) Which of the rivets shown will accurately fit the conical depression made by a 100° countersink?

1— A.
2— B.
3— C.
4— D.

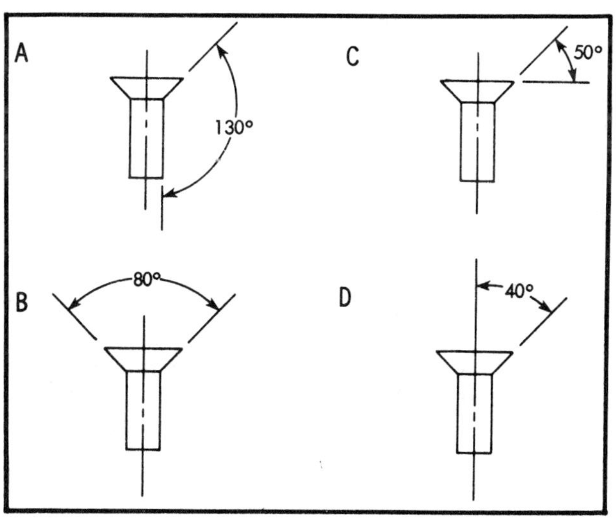

Figure 1

6113. Which is correct concerning the use of a file?

1— A file with an integral handle is referred to as a safe-edge file.
2— Apply pressure on the forward stroke, only, except when filing very soft metals such as lead or aluminum.
3— A smoother finish can be obtained by using a double-cut file than by using a single-cut file.
4— The terms "double-cut" and "second-cut" have the same meaning in reference to files.

6114. What is one of the determining factors which permits machine countersinking when flush riveting?

1— Thickness of the material and rivet diameter are the same.
2— Thickness of the material is less than the thickness of the rivet head.
3— Thickness of the material is greater than the thickness of the rivet head.
4— Thickness of the upset head of the rivet is greater than the material.

6115. When repairing a small hole on a metal stressed skin, the major consideration in the design of the patch should be

1— the shear strength of the riveted joint.
2— to use rivet spacing similar to a seam in the skin.
3— that the bond between the patch and the skin is sufficient to prevent dissimilar metal corrosion.
4— to use patch material one gauge thicker than the skin.

6116. Which procedure is correct when using a reamer to finish a drilled hole to the correct size?

1— Turn the reamer in the cutting direction when enlarging the hole and in the opposite direction to remove from the hole.
2— Turn the reamer only in the cutting direction.
3— Apply considerable pressure on the reamer when starting the cut and reduce the pressure when finishing the cut.
4— Use a high-viscosity cutting oil to lubricate the reamer.

6117. Repairs or splices involving stringers on the lower surface of stressed skin metal wings are usually

1—not permitted.
2—permitted but are normally more critical in reference to aerodynamic cleanness than similar repairs to the upper surface.
3—permitted only if the damage does not exceed 6 inches in any direction.
4—permitted but are normally more critical in reference to strength in tension than similar repairs to the upper surface.

6118. When straightening members made of 2024-T4, you should

1—apply heat to the outside of the bend.
2—straighten cold and reinforce.
3—straighten cold and anneal to remove stress.
4—apply heat to the inside of the bend.

6119. Clad aluminum alloys are used in aircraft because they

1—can be heat treated much easier than the other forms of aluminum.
2—are less subject to corrosion than uncoated aluminum alloys.
3—are stronger than unclad aluminum alloys.
4—are lighter than the other forms of aluminum.

6120. Which statement is true regarding a cantilever wing?

1—It employs lift wires instead of lift struts.
2—It has nonadjustable lift struts.
3—No external bracing is needed.
4—It requires only one lift strut on each side.

6121. Aircraft structural units, such as spars, engine supports, etc., which have been built up from sheet metal, are normally

1—repairable, using approved methods.
2—repairable, except when subjected to compressive loads.
3—not repairable, but must be replaced when damaged or deteriorated.
4—repairable, except when subjected to tensile loads.

6122. A factor which determines the minimum space between rivets is the

1—type of material being riveted.
2—length of the rivets being used.
3—diameter of the rivets being used.
4—thickness of the material being riveted.

6123. What should be the included angle of a twist drill for hard metal?

1—118°.
2—100°.
3—90°.
4—60°.

6124. When fabricating parts from alclad 2024-T3 aluminum sheet stock,

1—bends should be made with a small radius to develop maximum strength.
2—all bends must be 90° to the grain.
3—avoid excessive exposure to the moisture.
4—all scratches, kinks, tool marks, nicks, etc., must be held to a minimum.

6125. The purpose of a relief hole is to

1—lighten the metal.
2—check expansion.
3—relieve stress concentration and prevent cracking.
4—increase the strength.

6126. Monocoque fuselages derive their principal strength from

1—bulkheads and longerons.
2—longerons and formers.
3—the actual covering, metal or plywood.
4—metal stringers.

6127. Which part(s) of a semi-monocoque fuselage prevent(s) tension and compression from bending the fuselage?

1—Metal stringers.
2—The fuselage covering.
3—Longerons and stringers.
4—Bulkheads and skin.

6128. In making wing skin repairs,

1—rivet alloy is not a critical item.
2—2024-T rivets may be replaced with 2117-T rivets.
3—rivets selected should be the same alloy as the material being riveted.
4—2117-T rivets should be heat treated.

6129. Rivet pitch is the

1—distance between the centers of adjacent rivets in the same row.
2—distance from the center of the rivet to the edge of the repair.
3—countersunk head on flush rivets.
4—arc of a round head, brazier head, or universal head rivet.

6130. (Refer to figure 2.) Select the preferred sketch for proper countersinking.

1— All are acceptable.
2— B.
3— C.
4— A.

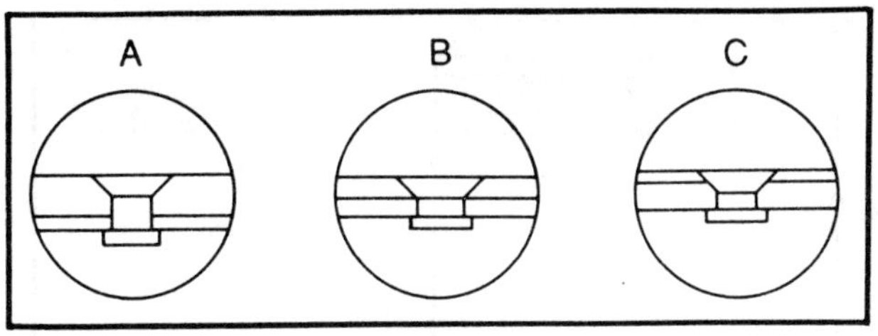

Figure 2

6131. What is usually indicated by a black "smoky" residue streaming back from some of the rivets on an aircraft?

1— The rivets are excessively tight.
2— The aircraft is being exposed to wide temperature variations in flight.
3— The aircraft is due to be washed.
4— Fretting corrosion between the rivets and the skin.

6132. The identifying marks on the heads of aluminum alloy rivets indicate the

1— corrosion proofing or other special process used in the manufacture of the rivets.
2— degree of dimensional and process control observed during manufacture.
3— head shape, shank size, material used, and specifications adhered to during manufacture.
4— specific alloy used in the manufacture of the rivets.

6133. The shear strength of an AN470DD-6-8 rivet, immediately after driving, is approximately 75 percent of its ultimate shear strength. The remaining 25 percent is obtained by

1— chilling the riveted unit for an hour.
2— standing at ordinary room temperature for about 4 days (age harden).
3— heating the unit in which the rivet was used to 960 °F, holding for 3 hours, and allowing the unit to cool slowly until room temperature is reached.
4— shear loads imposed on the riveted assembly in use.

6134. Which of the following need not be considered when determining minimum rivet spacing?

1— Rivet diameter.
2— Material thickness.
3— Rivet length.
4— Type of material being riveted.

6135. What is the purpose of refrigerating 2017 and 2024 aluminum alloy rivets after heat treatment?

1— To accelerate age hardening.
2— To prevent work hardening.
3— To relieve internal stresses.
4— To retard age hardening.

6136. Under certain conditions, type A rivets are not used because of their

1— low strength characteristics.
2— high alloy content.
3— high stress per unit area.
4— tendency toward embrittlement when subjected to vibration.

6137. A rivet set used to drive AN470 or MS20470 rivets should

1— have the same radius as the rivet head.
2— have a slightly greater radius than the rivet head.
3— be nearly flat on the end, with a slight radius on the edge to prevent damage to the sheet being riveted.
4— have a slightly smaller radius than the rivet head.

6138. Heat-treated rivets in the D and DD series that are not driven within the prescribed time after heat treatment or removal from refrigeration

1— must be reheat treated before use.
2— must be discarded.
3— may be returned to refrigeration and used later without reheat treatment.
4— may be reheat treated only one additional time.

6139. The dimensions of an AN430AD-4-8 rivet are

1— 4/32 inch in diameter and 8/32 inch long.
2— 4/16 inch in diameter and 8/16 inch long.
3— 1/8 inch in diameter and 1/2 inch long.
4— 4/16 inch in diameter and 8/32 inch long.

6140. The primary alloying agent of 2024-T36 is indicated by the number

1— 2.
2— 20.
3— 24.
4— 36.

6141. Which part of the 2017-T36 aluminum alloy designation indicates the primary alloying agent used in its manufacture?

1— 2.
2— 17.
3— 20.
4— 36.

6142. A sheet metal repair is to be made using two pieces of 0.040-inch aluminum riveted together. All rivet holes are drilled for 3/32-inch rivets. The length of the rivets to be used will be

1— 1/8 inch.
2— 7/32 inch.
3— 1/4 inch.
4— 5/16 inch.

6143. Most rivets used in aircraft construction have

1— two raised bars on their heads.
2— dimples.
3— smooth heads without markings.
4— a raised dot.

6144. AN426DD-6-5 indicates a countersunk rivet which has

1— a shank length of 5/16 inch (excluding head).
2— a shank length of 5/32 inch (excluding head).
3— an overall length of 5/16 inch.
4— a dimple in its head.

6145. Which rivet may be used as received without further treatment?

1— 2024-T4.
2— 2117-T3.
3— 2017-T3 over 3/16-inch diameter.
4— 2017-T31 over 3/16-inch diameter.

6146. (Refer to figure 3.) Which is the grip length of the flush rivet?

1— A.
2— B.
3— C.
4— D.

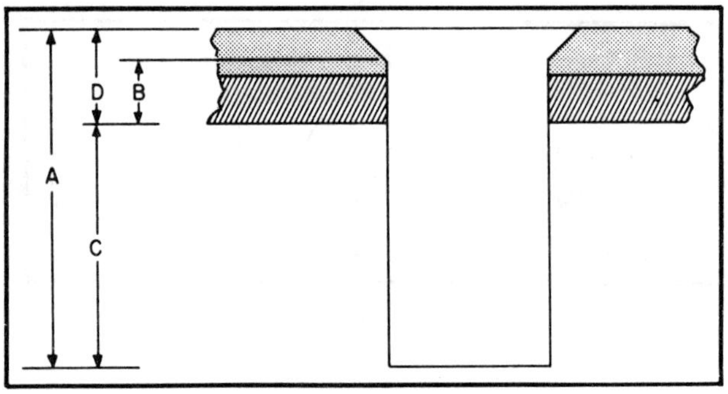

Figure 3

6147. Which rivets should be selected to join two sheets of .032 aluminum?

1— AN425D-4-3.
2— AN470AD-4-4.
3— AN455DD-5-3.
4— AN430-5-4.

6148. A sheet metal repair is to be made using two pieces of 0.0625-inch aluminum riveted together. All rivet holes are drilled for 1/8-inch rivets. The length of the rivets to be used will be

1— 1/8 inch.
2— 5/32 inch.
3— 3/16 inch.
4— 5/16 inch.

6149. Mild steel rivets are used for riveting

1— nickel-steel parts.
2— aluminum alloy parts.
3— magnesium parts.
4— steel parts.

6150. A DD rivet is heat treated before use to

1— harden and increase strength.
2— accelerate age hardening.
3— relieve internal stresses.
4— soften to facilitate riveting.

6151. When riveting dissimilar metals together, certain precautions must be exercised in order to prevent an electrolytic action. The best procedure to follow is to

1— treat the surfaces to be riveted together with a process called anodic treatment.
2— place a thin gasket of aluminum tape, cellophane, or properly impregnated fabric between the two parts.
3— paint the two surfaces as well as the surrounding area with a dope-proof paint.
4— avoid the use of dissimilar metals by redesigning the unit according to the recommendations outlined in AC 43.13-1A.

6152. The length of a rivet to be used to join a sheet of .032-inch and .064-inch aluminum alloy should be equal to

1— two times the rivet diameter plus .032 inch.
2— two times the rivet diameter plus .064 inch.
3— one and one-half times the rivet diameter plus .096 inch.
4— three times the rivet diameter plus .096 inch.

6153. What size drill is used to remove a rivet?

1— A drill two sizes smaller than the rivet shank diameter.
2— A drill one size smaller than the rivet shank diameter.
3— A drill the same size as the rivet shank diameter.
4— A drill two sizes larger than the rivet shank diameter.

6154. Joggles in removed rivet shanks would indicate partial

1— bearing failure.
2— torsion failure.
3— shear failure.
4— tear failure.

6155. What type loads cause the most rivet failures?

1— Shear.
2— Bearing.
3— Head.
4— Torsion.

6156. Which rivet is used for riveting magnesium alloy structures?

1— Mild steel.
2— 5056 aluminum.
3— Monel.
4— A1100 aluminum.

6157. Which rivet is used for riveting nickel-steel alloys?

1— 2017 aluminum.
2— 2024 aluminum.
3— Mild steel.
4— Monel.

6158. The length of rivet to be chosen when making a structural repair that involves the joining of 0.032 and 0.064 aluminum sheet, drilled with a No. 30 drill, is

1— 7/16 inch.
2— 7/32 inch.
3— 5/16 inch.
4— 1/4 inch.

6159. (Refer to figure 4.) The length of flat A is

1— 3.750 inches.
2— 3.875 inches.
3— 3.813 inches.
4— 3.937 inches.

6160. (Refer to figure 4.) The amount of material required to make the 90° bend is

1— 0.3434 inch.
2— 0.3717 inch.
3— 0.3925 inch.
4— 0.8397 inch.

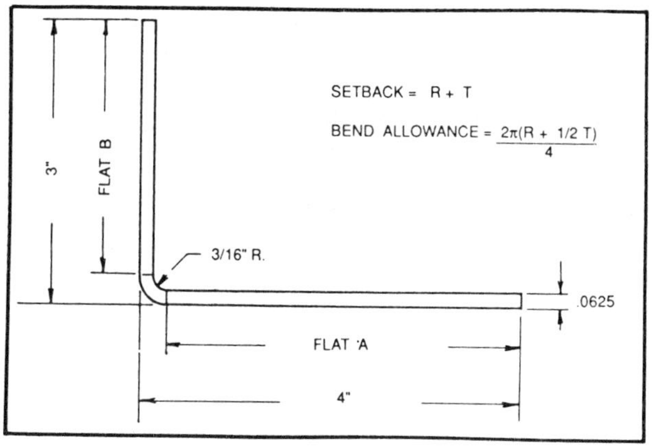

Figure 4

6161. (Refer to figure 5.) What is the length of flat A?

1—2.8 inches.
2—3.7 inches.
3—3.8 inches.
4—3.9 inches.

6162. (Refer to figure 5.) What is the flat layout dimension?

1—7.0 inches.
2—6.8 inches.
3—6.6 inches.
4—6.0 inches.

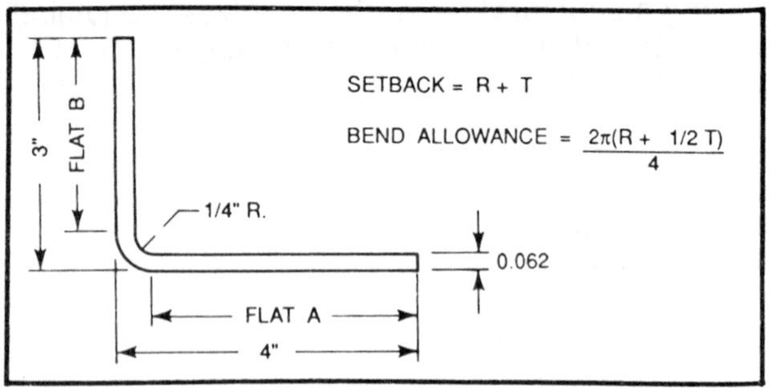

Figure 5

6163. If a streamline cover plate is to be hand formed using a form block, a piece of dead soft aluminum should first be placed over the hollow portion of the mold and securely fastened in place. The bumping operation should be

1—distributed evenly over the face of the aluminum at all times rather than being started at the edges or center.
2—started by tapping the aluminum lightly around the edges and gradually working down into the center.
3—started by tapping the aluminum in the center until it touches the bottom of the mold and then working out in all directions.
4—done alternately on the back and front of the metal to reduce internal stresses.

6164. A piece of flat stock that is to be bent to a closed angle of 15° must be bent through an angle of

1—165°.
2—105°.
3—75°.
4—15°.

6165. When a piece of aluminum alloy is to be bent using a minimum radius for the type and thickness of material,

1—the piece should be bent slowly to eliminate cracking.
2—the layout should be made so that the bend will be 90° to the grain of the sheet.
3—less pressure than usual should be applied with the movable (upper) clamping bar.
4—the movable (upper) clamping bar should be moved back from the working face of the bending leaf.

6166. The flat layout or blank length of a piece of metal from which a simple L-shaped bracket 3 inches by 1 inch is to be bent depends upon the radius of the desired bend. The bracket which will require the greatest amount of material is one which has a bend radius of

1—1/8 inch.
2—1/2 inch.
3—3/8 inch.
4—1/4 inch.

6167. If it is necessary to compute a bend allowance problem and bend allowance tables are not available, the neutral axis of the bend can be

1—represented by the actual length of the required material for the bend.
2—found by adding approximately one-half of the stock thickness to the bend radius.
3—found by subtracting the stock thickness from the bend radius.
4—that point at which the bent and unbent sections of one leg meet.

6168. Unless otherwise specified, the radius of a bend is the

1—inside radius of the metal being formed.
2—inside radius plus one-half the thickness of the metal being formed.
3—outside radius of the metal being formed.
4—radius of the neutral axis plus one-half the thickness of the metal being formed.

6169. The sharpest bend that can be placed in a piece of metal without critically weakening the part is called the

1—bend allowance.
2—minimum radius of bend.
3—maximum radius of bend.
4—neutral radius of bend.

6170. The most important factors needed to make a flat pattern layout are

1— radius, thickness, and mold line.
2— mold lines, mold point, and setback.
3— radius, thickness, and degree of bend.
4— the lengths of the legs (flat sections).

6171. A piece of sheet metal is bent to a certain radius. The curvature of the bend is referred to as the

1— bend line.
2— bend allowance.
3— neutral line.
4— bend radius.

6172. You can distinguish between aluminum and aluminum alloy by

1— filing the metal.
2— testing with vinegar.
3— testing with a 10 percent solution of caustic soda.
4— grinding and watching the sparks.

6173. What is the purpose of a joggle?

1— To allow clearance for a sheet or an extrusion.
2— To form a chamfered reinforcing ridge.
3— To increase obstruction for a sheet or an extrusion.
4— To decrease the weight of the part and still retain the necessary strength.

6174. When bending metal, the material on the outside of the curve stretches while the material on the inside of the curve compresses. That part of the material which is not affected by either stress is the

1— mold line.
2— setback line.
3— bend tangent line.
4— neutral line.

6175. (Refer to figure 6.) Determine the dimensions of A, B, and C in the flat layout.

Setback = .252
Bend allowance = .345

1— A = .748
B = 2.252
C = 2.004.
2— A = .748
B = 1.496
C = 1.248.
3— A = 1.252
B = 2.504
C = 1.752.
4— A = .655
B = 1.310
C = 1.845.

6176. (Refer to figure 6.) What is dimension D?

Setback = .252
Bend allowance = .345

1— 4.500.
2— 3.492.
3— 4.182.
4— 3.841.

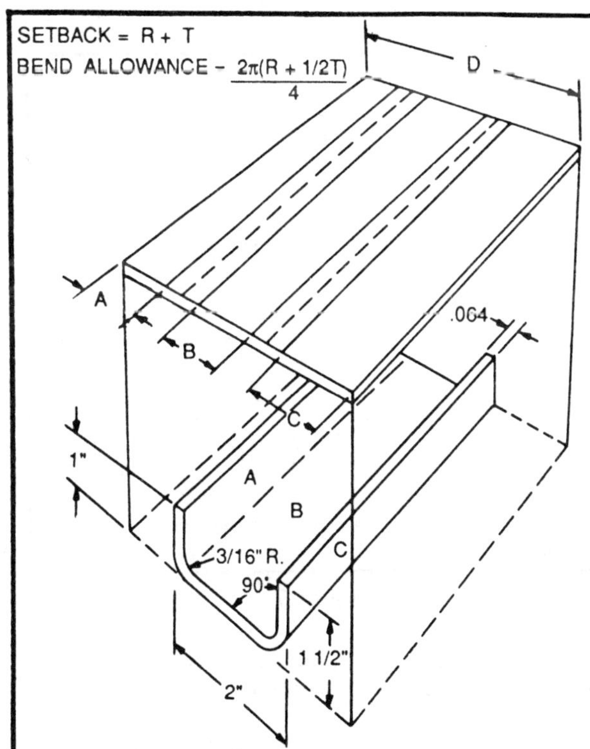

Figure 6

6177. When considering the location of the bend line for a fitting with only one bend,

1— add the bend radius to the bend tangent line of the smaller leg.
2— divide the bend allowance in half.
3— the setback must be known.
4— add the bend radius to the bend tangent line of the larger leg.

6178. Holes are usually drilled at the intersection of two bends to prevent the metal from cracking. These holes are referred to as

1— pilot holes.
2— relief holes.
3— lightening holes.
4— countersunk holes.

6179. (Refer to figure 7.) What is dimension F?

Setback at D = .095
Setback at E = .068
Bend allowance at D = .150
Bend allowance at E = .112

1— 6.000.
2— 4.836.
3— 5.936.
4— 5.738.

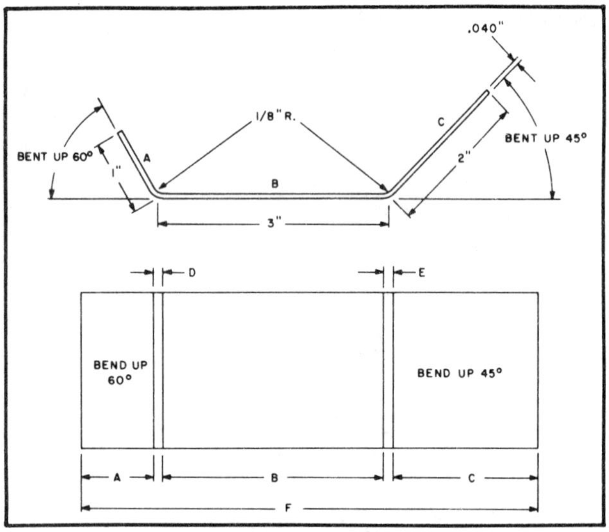

Figure 7

6180. On a sheet metal fitting layout with a single bend, allow for stretching by

1— adding the setback to one side only.
2— adding the setback to each leg.
3— subtracting the setback from one leg.
4— subtracting the setback from both legs.

6181. The aluminum alloys used in aircraft construction are usually hardened by which method?

1— Cold-working.
2— Aging.
3— Heat treatment.
4— Chemical.

6182. The primary reason for using flux when welding aluminum is to

1— prevent oxides from forming ahead of the weld.
2— promote better fusion of the base metal at a lower temperature.
3— prevent molten metal from flowing too widely.
4— clean the base metal ahead of the weld.

6183. Which statement best describes magnesium welding?

1— Magnesium can be welded to other metals.
2— Filler rod should be nickel-steel.
3— Use an oxidizing flame held at a flat angle to the work.
4— Filler rod should be the same composition as base metal.

6184. Magnesium aircraft structural members are usually heat treated. Which statement is true concerning the welding of heat-treated magnesium?

1— The welded section can never have the strength of the original metal.
2— Flux should not be used as it causes corrosion to commence.
3— Use an oxidizing flame held at a flat angle to the work.
4— Magnesium cannot be repaired by fusion welding.

6185. The oxyacetylene flame for silver soldering should be

1— harsh.
2— oxidizing.
3— neutral.
4— carburizing.

6186. Why is it necessary to use flux in all silver soldering operations?

1— To chemically clean the base metal of oxide film.
2— To prevent overheating of the base metal.
3— To increase the strength of the joint and save the expensive silver solder.
4— To increase heat conductivity.

6187. Engine mount members should preferably be repaired by using a

1— larger diameter tube with fishmouth and no rosette welds.
2— larger diameter tube with fishmouth and rosette welds.
3— smaller diameter tube with fishmouth and rosette welds.
4— larger diameter tube with 30° cuts and rosette welds.

6188. What method of repair is recommended for a steel tube longeron dented at a cluster?

1— Welded split sleeve.
2— Welded outer sleeve.
3— Welded patch plate.
4— Welded inner sleeve.

6189. Welding over a previously brazed or soldered joint is

1— not permitted.
2— permitted only if welding temperatures do not exceed 2,000 °F.
3— permitted only if all traces of the braze or solder material are removed by chemical treatment.
4— permitted only if done within a carefully controlled reducing atmosphere.

6190. Which statement concerning soldering is correct?

1— Joints in electric wire to be soldered should be mechanically secure prior to soldering.
2— Changeable shades of blue can be observed on the surface of a copper soldering tip when the proper temperature for soldering has been reached.
3— If the soldering temperature is too high, the solder will form in lumps and not produce a positive bond.
4— If soldering temperatures are too low, the solder will flow well but will have a tendency to follow the soldering tip.

6191. When a wire or cable is to be permanently attached to an electrical component by soldering, the

1— wire or cable should be thoroughly tinned prior to attachment to increase its flexibility.
2— joint should be mechanically secure before soldering.
3— joint should be able to depend upon the solder for its mechanical strength.
4— solder should be made to coat the wire completely and smoothly and not penetrate between the wire strands.

6192. A resurfaced soldering iron cannot be used effectively until after the working face has been

1— carbonized.
2— fluxed.
3— roughened.
4— tinned.

6193. Which steel parts are normally considered repairable by welding?

1— Brazed or soldered parts.
2— Turnbuckle ends.
3— Streamline wire braces.
4— SAE 4130 chrome/molybdenum tubing.

6194. In aircraft welding, the usual practice in the selection of a welding tip is to use

1— the type of material to be welded as the primary basis for selection.
2— a tip with a hole size equal to the diameter of the welding rod used.
3— as small a tip as possible with the tip adjusted to the maximum operating range.
4— as large a tip as possible with the tip adjusted to the maximum operating range.

6195. Why should a carburizing flame be avoided when welding steel?

1— It removes the carbon content.
2— It hardens the surface.
3— It causes excessive sparking.
4— A cold weld will result.

6196. Which item is the most important consideration when selecting a welding rod?

1— Type of torch.
2— Type of metal to be welded.
3— Thickness of the metal to be welded.
4— Regulator pressure.

6197. When welding a tank fabricated from .071 sheet aluminum, a welding tip slightly larger than that used for welding steel of the same thickness should be used to obtain sufficient heat to melt the base metal. The flame used for aluminum welding should

1— be neutral and soft.
2— be slightly oxidizing.
3— contain an excess of acetylene and leave the tip at a relatively low speed.
4— contain an excess of acetylene and leave the tip at a relatively high speed.

6198. Why should a new or resurfaced soldering copper tip be tinned or coated with solder?

1—To aid in the transfer of heat from the soldering tip to the joint to be soldered.
2—So that the solder on the tip will transfer to the joint to be soldered as soon as the proper temperature is reached.
3—To prevent the soldering tip from overheating the joint to be soldered.
4—To prevent excessive heat from being radiated from the tip.

6199. A very thin and pointed tip on a soldering copper is undesirable because it will

1—burn the alloys out of the solder.
2—be very difficult to tin.
3—cool too rapidly.
4—punch holes in the metal being soldered.

6200. A gas weld which is successfully completed should have which characteristics?

1—The finish weld should have a rough seam and be nonuniform in thickness.
2—The weld metal should be tapered smoothly into the base metal.
3—Oxide should be formed on the base metal close to the weld.
4—The base metal should show signs of pits.

6201. An acceptable line pressure for acetylene is

1—5 PSI.
2—18 PSI.
3—22 PSI.
4—the same as the oxygen pressure.

6202. Cylinders used to transport and store acetylene

1—are purged after each use.
2—are pressure tested to 3,000 PSI.
3—are green in color.
4—contain acetone.

6203. A welding torch backfire may be caused by

1—a loose tip.
2—moving the torch too rapidly across the welding surface.
3—using too much acetylene.
4—an underheated tip.

6204. Which statement concerning a welding process is true?

1—The inert-arc welding process uses an inert gas to protect the weld zone from the atmosphere.
2—In the metallic-arc welding process, filler material, if needed, is provided by a separate metal rod of the proper material held in the arc.
3—In the carbon-arc welding process, filler material, if needed, is provided by the carbon rod.
4—In the oxyacetylene welding process, the filler rod used for steel is covered with a thin coating of flux.

6205. Where should the flux be applied when oxyacetylene welding aluminum?

1—Painted only on the surface to be welded.
2—Painted on the surface to be welded and applied to the welding rod.
3—Applied only to the welding rod.
4—Painted on the surface to be welded and applied to the welding rod after tack weld.

6206. What purpose does flux serve in welding aluminum?

1—Removes dirt, grease, and oil.
2—Minimizes or prevents oxidation.
3—Ensures proper distribution of the filler rod.
4—Aids in the fusing of the base metal but not the filler rod.

6207. Why are aluminum plates one-fourth inch or more thick usually preheated before welding?

1—Reduces internal stresses and assures more complete penetration.
2—Reduces welding time.
3—Prevents corrosion and ensures proper distribution of flux.
4—Removes oxide, dirt, grease, and oil.

6208. How should a welding torch flame be adjusted to weld stainless steel?

1—Slightly carburizing.
2—Slightly oxidizing.
3—Neutral.
4—Heavy.

6209. Oxides form very rapidly when alloys or metals are hot. It is important, therefore, when welding aluminum to use a

1—solvent.
2—float.
3—filler.
4—flux.

6210. In gas welding, the amount of heat applied to the material being welded is controlled by the

1— amount of gas pressure used.
2— size of the tip opening.
3— distance the tip is held from the work.
4— temperature of the flame.

6211. Oxygen and acetylene cylinders are made of

1— heat-treated seamless copper.
2— seamless aluminum.
3— steel.
4— bronze.

6212. The maximum working pressure of acetylene when welding should never exceed

1— 15 PSI.
2— 18 PSI.
3— 25 PSI.
4— 29 PSI.

6213. When inspecting a butt-welded joint by visual means,

1— the penetration should be 25 to 50 percent of the thickness of the base metal.
2— the width of the bead should be twice the thickness of the base metal.
3— the penetration should be 100 percent of the thickness of the base metal.
4— look for evidence of excessive heat in the form of a very high bead.

6214. Annealing of aluminum

1— increases the tensile strength.
2— makes the material brittle.
3— removes stresses caused by forming.
4— makes the material hard.

6215. After welding aluminum, it is important that all traces of flux be removed. If left on the weld, it will

1— remove the alodizing.
2— cause paint to discolor and stain.
3— inhibit inspection of the weld.
4— corrode the metal.

6216. Edge notching is generally recommended in butt welding above a certain thickness of aluminum because it

1— helps hold the metal in alignment during welding.
2— results in a smoother bead.
3— makes the removal of flux traces easier afterward.
4— aids in getting full penetration and prevents local distortion.

6217. If too much acetylene is used in the welding of stainless steel,

1— the metal will bubble and a porous weld will result.
2— the metal will absorb carbon and lose its resistance to corrosion.
3— oxide will be formed on the base metal close to the weld.
4— the base metal will show signs of burns, pits, cracks, or distortion.

6218. The tail rotor of a helicopter could fail to compensate accurately for the torque of the main rotor if the

1— power transmission is erratic.
2— engine power is reduced.
3— rotor is inaccurately rigged.
4— engine power is applied suddenly to the main rotor.

6219. The vertical flight of a helicopter is controlled by

1— increasing or decreasing collective pitch.
2— tilting the rotor disk.
3— cyclic pitch changes.
4— increasing or decreasing the RPM of the main rotor.

6220. A decrease in pitch angle of the tail rotor blades on a helicopter

1— causes the tail to pivot in the opposite direction of torque rotation around the main rotor axis.
2— is produced by depressing the left anti-torque pedal.
3— causes the tail to pivot in the direction of torque rotation around the main rotor axis.
4— is required to counteract main rotor torque produced by takeoff RPM.

6221. Tracking of helicopter rotor blades is normally done

1—on a special fixture before they are installed due to their flexibility.
2—while the blades are rotating at a specified RPM and collective pitch angle.
3—by triangulation while the blades are at rest and in a specified location in reference to the fuselage centerline.
4—by direct measurement from the fuselage structure to a specified blade station when the blade is parallel to the fuselage centerline.

6222. The acute angle formed by the chord line of a wing and the relative wind is known as the

1—longitudinal dihedral angle.
2—angle of climb.
3—angle of incidence.
4—angle of attack.

6223. A helicopter in forward flight, cruise configuration, changes direction by

1—varying the pitch of the main rotor blades.
2—changing tail rotor RPM.
3—tilting the main rotor disk in the desired direction.
4—tilting the tail rotor.

6224. If a severe abnormal vertical vibration developed in a helicopter equipped with a two-blade rotor system, this would indicate the

1—rotor blades are out of balance.
2—blade coning stops have failed.
3—blade droop stops are not adjusted properly.
4—rotor blades are out of track.

6225. In a hovering helicopter equipped with a tail rotor, directional control is maintained by

1—changing the tail rotor RPM.
2—tilting the main rotor disk in the desired direction.
3—a conventional rudder and rudder control system.
4—varying the pitch of the tail rotor blades.

6226. If a single-rotor helicopter is in horizontal flight, the angle of attack of the advancing blade is

1—more than the retreating blade.
2—equal to the retreating blade.
3—the same at any point around the rotor disk.
4—less than the retreating blade.

6227. Greater downwash effect through the aft part of the main rotor disk compared to the forward part of the disk is

1—Coriolis effect.
2—transverse flow lift.
3—translating tendency.
4—transitional lift.

6228. Main rotor blades that do not cone by the same amount during rotation are said to be out of

1—balance.
2—lateral cyclic pitch.
3—collective pitch.
4—track.

6229. One purpose of the clutch provided between the engine and the helicopter transmission is to

1—disconnect the rotor from the engine to relieve the starter load.
2—automatically disengage the rotor from the engine in case of an engine failure.
3—permit practice of autorotation landings.
4—permit the rotor to operate at a greater RPM than provided for by maximum engine RPM.

6230. Which statement is correct concerning torque effect on helicopters?

1—As horsepower increases, torque decreases.
2—Torque direction is the same as rotor blade rotation.
3—As horsepower decreases, torque increases.
4—Torque direction is the opposite of rotor blade rotation.

6231. How is the torque force associated with single-rotor helicopter compensated?

1—A tail rotor with a variable pitch mechanism that is actuated by pilot controls.
2—A twist in the main rotor blade chord.
3—A vertical flat plate that is acted upon by the main rotor downwash.
4—A double set of planetary gears in the main transmission.

6232. What is the purpose of the free-wheeling unit in a helicopter drive system?

1—It disconnects the rotor whenever the engine stops or slows below the equivalent of rotor RPM.
2—It releases the rotor brake for starting.
3—It relieves bending stress on the rotor blades during starting.
4—It allows the engine to be over-revved for landing.

6233. Movement about the longitudinal axis (roll) in a helicopter is effected by movement of

1—the drag hinge damper control.
2—the collective pitch control.
3—the cyclic pitch control.
4—the tail rotor pitch control.

6234. Movement about the lateral axis (pitch) in a helicopter is effected by movement of

1—the drag hinge damper control.
2—the collective pitch control.
3—the cyclic pitch control.
4—the tail rotor pitch control.

6235. Lateral dihedral, a rigging consideration on most airplanes of conventional design, contributes most to stability of the airplane about its

1—longitudinal axis.
2—vertical axis.
3—lateral axis.
4—transverse axis.

6236. As the result of an accident, it is necessary to replace the fuselage fittings for the wing attachment. Prior to welding the fittings in place, a great deal of care should be exercised in locating them in their proper position because the

1—angle of attack will be re-established at this time.
2—angle of incidence is one of the factors to be re-established at this time.
3—main spar attachment fittings must line up with fuselage fittings; otherwise, the spar location in the wing will have to be changed.
4—aspect ratio of the wing will be controlled by the fuselage fittings.

6237. If a pilot reports that an airplane flies left wing heavy, this condition is corrected by

1—washing-in the right wing, or washing-out the left wing, or both.
2—washing-in the left wing, or washing-out the right wing, or both.
3—washing-in the right wing only.
4—washing-out the left wing only.

6238. If the vertical fin of a single-engine, propeller-driven airplane is rigged properly, it will generally be parallel to

1—the longitudinal axis but not the vertical axis.
2—the vertical axis but not the longitudinal axis.
3—neither the longitudinal nor the vertical axis.
4—both the longitudinal and vertical axes.

6239. An airplane which has good longitudinal stability should have a minimum tendency to

1—roll.
2—pitch.
3—yaw.
4—stall.

6240. As the angle of attack of an airfoil increases, the center of pressure will

1—move toward the trailing edge.
2—remain stationary because both lift and drag components increase proportionally to increased angle of attack.
3—remain stationary because of no change in the incidence angle.
4—move toward the leading edge.

6241. The angle of incidence is that acute angle formed by

1—the angular difference between the setting of the main airfoil and the auxiliary airfoil (horizontal stabilizer) in reference to the longitudinal axis of the aircraft.
2—the angular difference between the wing settings of the two wings of a biplane.
3—a line parallel to the wing chord and a line parallel to the longitudinal axis of the aircraft.
4—a line parallel to the wing from root to tip and a line parallel to the lateral axis of the aircraft.

6242. An airplane's center of lift is usually located aft of its center of gravity

1—so that the airplane will have a tail-heavy tendency.
2—so that the airplane will have a nose-heavy tendency.
3—since it is impossible to predict the exact location of either, it is considered next best to having them both fall at the same point.
4—to improve stability about the longitudinal axis.

6243. An airplane is controlled directionally about its vertical axis by

1—the rudder.
2—the elevator(s).
3—the ailerons.
4—a combination of two of the above.

6244. The elevators of a conventional airplane are used to provide rotation about the

1—longitudinal axis.
2—lateral axis.
3—directional axis.
4—vertical axis.

6245. Washing-in the left wing of a monoplane, for purposes of rigging corrections after flight test, will have what effect on the lift and drag of that wing?

1—Both drag and lift will decrease due to decreased angle of attack.
2—Both drag and lift will increase due to increased angle of attack.
3—The lift will decrease due to the effect of the drag increase.
4—The drag will decrease due to the effect of the lift increase.

6246. Flaps increase the effective lift of an airfoil by

1—increasing the camber of the airfoil.
2—introducing drag aft of the center of pressure.
3—reducing the profile drag.
4—increasing the angle of attack of the airfoil.

6247. If the right wing of a monoplane is improperly rigged to a greater angle of incidence than designated in the manufacturer's specifications, it will cause the

1—airplane to be off balance both laterally and directionally.
2—airplane to pitch and roll about the lateral axis.
3—airplane to be out of lateral balance only.
4—right wing to have both an increased lift and a decreased drag.

6248. The chord of a wing is measured from

1—wingtip to wingtip.
2—wing attachment point to the wingtip.
3—leading edge to trailing edge.
4—maximum upper camber to the base line.

6249. When the lift of an airfoil increases, the drag will

1—decrease.
2—not be affected.
3—also increase.
4—increase while the lift is changing but will return to its original value.

6250. What physical factors are involved in the aspect ratio of airplane wings?

1—Thickness and chord.
2—Span and chord.
3—Dihedral and angle of attack.
4—Sweepback and lateral axis.

6251. Improper rigging of the elevator trim tab system will affect the balance of the airplane about its

1—lateral axis.
2—longitudinal axis.
3—vertical axis.
4—yawing moment.

6252. An airplane that has a tendency to gradually increase a pitching moment that has been set into motion has

1—poor longitudinal stability.
2—good lateral stability.
3—poor lateral stability.
4—good longitudinal stability.

6253. A wing slat is a movable airfoil attached to the leading edges of high-performance airplane wings. Their purpose is to

1—reduce stalling speed.
2—replace flaps.
3—act as a dive brake or speed brake.
4—increase speed on takeoff.

6254. The angle of incidence of an airplane

1—is changed by the pilot while ascending.
2—affects the dihedral of the wings.
3—is that angle between the relative wind and the chord of the wing.
4—does not change in flight.

6255. Buffeting is the intermittent application of forces to a part of an airplane. It is caused by

1—incorrect rigging of flaps.
2—an unsteady flow from turbulence.
3—incorrect rigging of ailerons.
4—unknown forces.

6256. Movement of an airplane along its lateral axis (roll) is also movement

1—around or about the vertical axis controlled by the rudder.
2—around or about the longitudinal axis controlled by the elevator.
3—around or about the lateral axis controlled by the ailerons.
4—around or about the longitudinal axis controlled by the ailerons.

6257. An airplane should be rigged to fly hands-off at

1—cruising speed.
2—sea level.
3—never-exceed speed.
4—landing speed.

6258. The primary purpose of stall strips is to

1—provide added lift at slow speeds.
2—stall the inboard portions of the wings first.
3—provide lateral stability at cruise speed and above.
4—provide added lift at high angles of attack.

6259. Rigging and alignment checks should not be undertaken in the open; however, if this cannot be avoided, the aircraft should be positioned

1—with the tail into the wind.
2—makes no difference if the wind is steady (not gusting).
3—only in an enclosed hangar. These checks are never permissible in the open.
4—with the nose into the wind.

6260. The correct dihedral angle can be determined by

1—measuring the angular setting of each wing at the rear spar with a bubble protractor.
2—using a steel tape to measure the distance of the wingtips and wing roots from the floor.
3—placing a straightedge and bubble protractor across the spars while the airplane is in flying position.
4—using a dihedral board and bubble level along the front spar of each wing.

6261. The dihedral angle of a wing may be measured by placing a straightedge and level protractor on the

1—front spar.
2—rear spar.
3—wing root.
4—wing chord.

6262. Where would you find precise information to perform a symmetry alignment check for a particular aircraft?

1—Aircraft Specification or Type Certificate Data Sheet.
2—Manufacturer's service bulletins.
3—AC 43.13-1A.
4—Aircraft service or maintenance manual.

6263. Where is the buttock line or buttline of an aircraft?

1—A height measurement left or right of, and perpendicular to, the horizontal centerline.
2—A width measurement left of, and perpendicular to, the vertical centerline.
3—A width measurement left or right of, and parallel to, the vertical centerline.
4—A height measurement right of, and parallel to, the horizontal centerline.

6264. Where is fuselage station No. 137 located?

1—137 centimeters aft of the nose or fixed reference line.
2—137 inches aft of the zero or fixed reference line.
3—Aft of the buttock line.
4—137 inches forward of the empennage.

6265. Proper wing twist in a sheet metal constructed wing can usually be checked by utilizing a

1—plum bob, string, and straightedge.
2—bubble level and special fixtures described by the manufacturer.
3—straightedge, tape measure, and carpenter's square.
4—carpenter's square and prop protractor.

6266. The control cable terminals on most late model aircraft are swaged and a painted band is placed around the cable adjacent to the terminal in order to

1—ascertain if the cable is safetied properly.
2—protect the cable and the fitting from electrolytic corrosion.
3—disclose twisting of the cable in the fitting.
4—detect slippage of the cable in the fitting.

6267. What nondestructive checking method is normally used to ensure that the correct amount of swaging has taken place when installing swaged-type terminals on aircraft control cable?

1—Magnaflux the completed assembly.
2—Measure the finished length of the terminal barrel and compare with the beginning length.
3—Use an after gauge to check the diameter of the swaged portion of the terminal.
4—Check the surface of the swaged portion of the terminal for cracks which indicate incomplete swaging.

6268. When inspecting a control cable turnbuckle for proper installation, determine that

1—no more than four threads are exposed on either side of the turnbuckle barrel.
2—the terminal end threads are visible through the safety hole in the barrel.
3—the safety wire ends are wrapped a minimum of four turns around the terminal end shanks.
4—only stainless steel safety wire has been used for safetying.

6269. If all instructions issued by the swaging tool manufacturer are followed when swaging a cable terminal, the resultant swaged terminal strength should be

1— the full rated strength of the cable.
2— 80 percent of the full rated strength of the cable.
3— 70 percent of the full rated strength of the cable.
4— 50 percent of the full rated strength of the cable.

6270. Which is an acceptable safety device for a castle nut when installed on secondary structures?

1— Fiber insert.
2— Star washer.
3— Lockwasher.
4— Cotter pin.

6271. When used in close proximity to magnetic compasses, cotter pins are made of what material?

1— Corrosion resisting steel.
2— Anodized aluminum alloy.
3— Magnesium and titanium alloy.
4— Cadmium-plated low carbon steel.

6272. When a fiber stop nut can be threaded on a bolt or stud through the fiber with only the fingers, it should be

1— torqued twice.
2— rejected.
3— reused once.
4— replated.

6273. The purpose of the vertical fin is to provide

1— vertical stability.
2— directional stability.
3— longitudinal stability.
4— lateral stability.

6274. How are changes in direction of a control cable accomplished?

1— Pulleys.
2— Ferrules.
3— Bell cranks.
4— Fairleads.

6275. What is the smallest size cable that may be used in aircraft primary control systems?

1— 3/16 inch.
2— 1/4 inch.
3— 5/16 inch.
4— 1/8 inch.

6276. After repairing or re-covering a rudder,

1— the surface should be rebalanced to its spanwise axis.
2— the surface should be rebalanced in its normal flight position.
3— the surface should be rebalanced to manufacturer's specifications.
4— rebalancing is not necessary due to its vertical position.

6277. A satisfactory method of inspecting control cables for broken wires is to

1— place a piece of cloth around the cable and run it back and forth over the length of the cable.
2— closely check the cable at all pulleys and fairleads with a 10-power magnifying glass.
3— remove the cable and check its entire length by the magnetic particle inspection method.
4— run a small permanent magnet over the cable. Broken wires will be pulled out of the cable making them easy to find by visual inspection.

6278. The cable-operated control system of an all-metal aircraft, not incorporating a temperature compensating device, has been rigged to the correct tension in a heated hangar. If the aircraft is operated in very cold weather, the cable tension will

1— decrease when the aircraft structure and cables become cold.
2— increase when the aircraft structure and cables become cold.
3— be unaffected if tin-plated high carbon steel cable is installed.
4— be unaffected if stainless steel pre-stretched cable is installed.

6279. Why should a control surface with counterweights be statically balanced?

1— To hold the surface streamlined when on the ground.
2— To make a control surface tab unnecessary.
3— To make installation and rigging easier.
4— To reduce the possibility of control surface flutter.

6280. During the inspection of a cable control system, a pulley would not rotate. Further inspection revealed a flat spot in the cable groove. What procedure should be followed?

1— The pulley should be replaced.
2— The cable groove should be reworked.
3— The pulley should be freed.
4— The pulley should be turned 90°.

6281. Fairleads should never deflect the alignment of a cable more than

1— 45°.
2— 12°.
3— 8°.
4— 3°.

6282. When installing a control cable, it is found that one end of a splice comes within 1/2 inch of entering a pulley groove with the loaded cable at the end of its travel. This condition

1— can be corrected by using swaged terminals on the replacement cable.
2— can be corrected by installing a smaller diameter pulley.
3— must be corrected by adjustment or replacement.
4— does not require further adjustment or replacement.

6283. With which system is differential control associated?

1— Trim system.
2— Aileron system.
3— Rudder system.
4— Elevator system.

6284. Which statement concerning the 100-hour inspection of an airplane equipped with a push-pull tube-type control system is true?

1— The threaded rod ends should be checked to determine that the ball bearing end is properly safetied to the push-pull rod with brass or stainless steel safety wire.
2— The threaded rod ends should not be adjusted in length for rigging purposes because the rod ends have been properly positioned and staked during manufacture.
3— The terminal end threads of the turnbuckles should be visible through the safety hole in the barrel.
4— The threaded rod ends should be checked for the amount of thread engagement by means of the inspection hole provided.

6285. If control cables are adjusted properly and the control surfaces tend to vibrate, the probable cause is

1— improper loading of the aircraft.
2— worn attachment fittings.
3— oil can effects on the control surfaces.
4— excessive cable tension.

6286. Aircraft flight control trim systems must be designed and installed so that the

1— pilot can determine the relative position of the trim tab from the cockpit.
2— operating control and trim tab will always move in the same direction.
3— operating control and the trim tab will always move in opposite directions.
4— trim system will disengage or become inoperative if the primary flight control system fails.

6287. Stability about the axis which runs parallel to the line of flight is referred to as

1— longitudinal stability.
2— lateral stability.
3— dynamic stability.
4— directional stability.

6288. The purpose of spring tabs or servo-tabs is to

1— assist the pilot in moving the control surfaces.
2— contribute to the static balance of the control surface.
3— balance the weight of that portion of the control surface located aft of the hinge line.
4— make in-flight trim adjustments possible.

6289. If the control stick of an aircraft with properly rigged flight controls is moved rearward and to the left, the right aileron will move

1— down and the elevator will move down.
2— up and the elevator will move down.
3— up and the elevator will move up.
4— down and the elevator will move up.

6290. Movement of the cockpit control toward the nosedown position during a ground operational check of the elevator trim tab system will cause the trailing edge of the trim tab to move in which direction?

1— Downward regardless of elevator position.
2— Upward regardless of elevator position.
3— Downward if the elevator is in the UP position and upward if the elevator is in the DOWN position.
4— Upward if the elevator is in the UP position and downward if the elevator is in the DOWN position.

6291. If the control stick of an aircraft with properly rigged flight controls is moved forward and to the right, the left aileron will move

1— up and the elevator will move down.
2— up and the elevator will move up.
3— down and the elevator will move up.
4— down and the elevator will move down.

6292. If the travel of an airplane's controls is correct but the cables are rigged exceptionally tight, what probable effect will this have when flying the airplane?

1— The airplane will tend to fall off on one wing.
2— The airplane will be heavy on the controls.
3— The pilot will be unable to fly the airplane hands-off.
4— The pilot will be unable to control the airplane in flight.

6293. During inspection of the flight control system of an airplane equipped with differential-type aileron control, side-to-side movement of the control stick will cause

1— each aileron to have a greater up travel (from the streamlined position) than down travel.
2— each aileron to have greater down travel (from the streamlined position) than up travel.
3— the right aileron to move through a greater number of degrees (from full up to full down) than the left aileron.
4— the left aileron to move through a greater number of degrees (from full up to full down) than the right aileron.

6294. A universal propeller protractor used to measure the degrees of aileron travel should be zeroed

1— before placing the protractor on the aircraft.
2— with the aileron in the NEUTRAL position.
3— with the aileron in the DOWN position.
4— when the aircraft is in a level flight attitude.

6295. The universal propeller protractor can be used to measure

1— propeller track.
2— inches of propeller blade travel.
3— aspect ratio of a wing.
4— degrees of flap travel.

6296. (Refer to figure 8.) Identify the cable that is used in primary control systems and in other places where operation over pulleys is frequent.

1— A.
2— B.
3— C.
4— D.

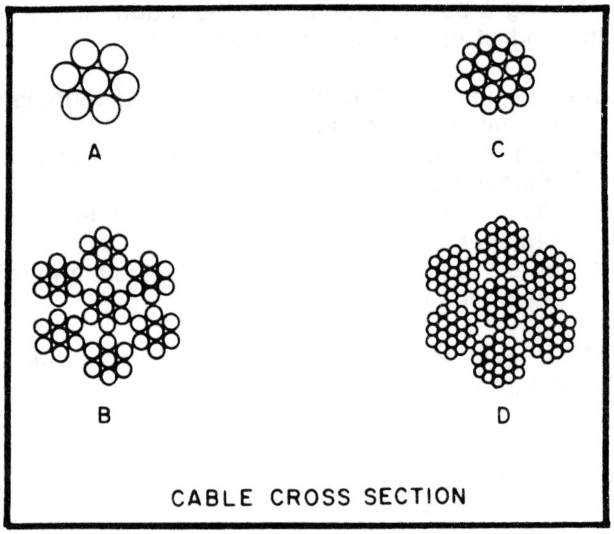

Figure 8

6297. Why are counterweights incorporated in the leading edge of primary control surfaces?

1— To enhance rapid movement of the surfaces.
2— To make the surfaces harder to move in flight.
3— To prevent rapid movement of surfaces during aerobatic maneuvers.
4— To prevent the surfaces from fluttering during flight.

6298. A tension regulator in the flight control cable system of a large all-metal aircraft is used primarily to

1— decrease the cable tension in cold weather.
2— increase the cable tension in cold weather.
3— provide a means of changing cable tension in flight.
4— retain a set tension.

6299. (Refer to figure 9.) When the outside air temperature is 80 °F, select the acceptable 3/16 cable tension range.

1— 115 pounds minimum, 143 pounds maximum.
2— 130 pounds minimum, 140 pounds maximum.
3— 117 pounds minimum, 143 pounds maximum.
4— 120 pounds minimum, 140 pounds maximum.

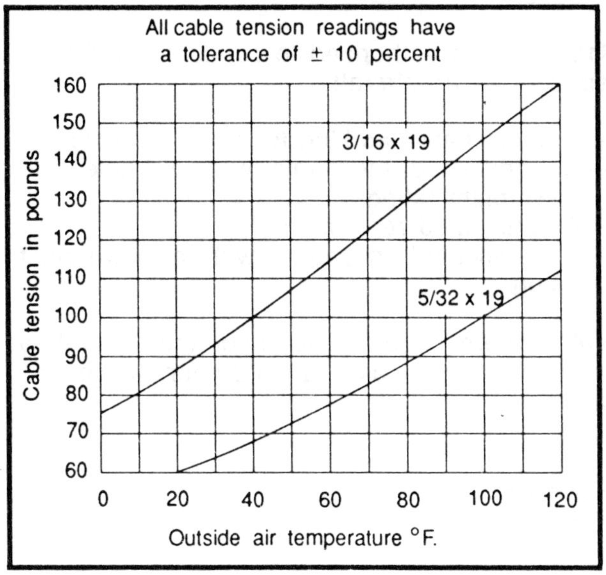

Figure 9

6300. Differential control on an aileron system means that

1— the down travel is more than the up travel.
2— the up travel is more than the down travel.
3— they travel up or down equally, but bell cranks are used.
4— one aileron on one wing travels further up than the aileron on the opposite wing to adjust for wash-in and wash-out.

6301. Why is it generally necessary to jack an aircraft indoors for weighing?

1— So aircraft may be placed in a level position.
2— To determine aircraft empty weight.
3— To stabilize the weighing scales.
4— So weighing scales may be calibrated to 0 pounds.

6302. Which should be accomplished before jacking an aircraft?

1— Remove all optional equipment.
2— Install critical stress panels or plates.
3— Determine that the fuel tanks are empty.
4— Make sure the aircraft is leveled laterally.

6303. Which statement about Airworthiness Directives (AD's) is true?

1— AD's can be complied with only by the manufacturer of the airframe, powerplant, or component involved.
2— AD's are information bulletins issued by the airframe, powerplant, or component manufacturer.
3— Compliance with an AD is not mandatory unless the aircraft affected is for hire.
4— Compliance with an applicable AD is mandatory and must be recorded in the permanent maintenance records.

6304. When overhauling electrical equipment, all necessary information should be obtained from

1— the applicable sections of AC 43.13-1A.
2— maintenance instructions published by the aircraft and/or equipment manufacturer.
3— a certificated airframe and/or powerplant mechanic.
4— Aircraft Specifications and/or Type Certificate Data Sheets.

6305. Which statement is correct when an aircraft is not approved for return to service after an annual inspection because of an item requiring major repair?

1— A certificated repair station may repair the defect, but an appropriately rated mechanic must approve the aircraft for return to service.
2— An appropriately rated mechanic may repair the defect, and an IA may approve the aircraft for return to service.
3— An appropriately rated mechanic may repair the defect and approve the aircraft for return to service.
4— Only the person who performed the annual inspection may approve the aircraft for return to service.

6306. Radio equipment installations made in accordance with Supplemental Type Certificate data require

1— approval using minor alteration procedures.
2— approval by an airframe and powerplant mechanic.
3— approval using major alteration procedures.
4— submission of engineering design data for each approval.

6307. When can an aircraft be operated with a 100-hour inspection overdue?

1—If necessary to reach a place at which the inspection can be done.
2—Only when the aircraft is operated under a continuous airworthiness program.
3—When the aircraft is operated under a progressive inspection program.
4—If aircraft is equipped with an hour meter.

6308. Where would you find the recommended statement for recording the approval or disapproval for return to service of an aircraft after a 100-hour or annual inspection?

1—FAR Part 36.
2—FAR Part 65.
3—FAR Part 43.
4—FAR Part 91.

6309. The maximum time a 100-hour inspection may be extended is

1—10 hours.
2—2 hours.
3—5 hours.
4—15 hours.

6310. Which statement is correct when an aircraft has not been approved for return to service after an annual inspection because of several items requiring minor repair?

1—Only the person who performed the annual inspection may approve the aircraft for return to service.
2—An appropriately rated mechanic may repair the defects and approve the aircraft for return to service.
3—An appropriately rated mechanic may repair the defects, but an IA must approve the aircraft for return to service.
4—An authorized repair station may repair the defects, but an appropriately rated mechanic must approve the aircraft for return to service.

6311. An aircraft that is due an annual inspection may be operated

1—no more than 50 nautical miles from its point of origin.
2—if a special permit has been issued for the aircraft.
3—for the purpose of performing maintenance.
4—for a period of time not to exceed 10 hours.

6312. Which is a portion of the required permanent maintenance record entry after a 100-hour inspection?

1—The name of the owner of the aircraft.
2—The date of the inspection and aircraft time in service.
3—A brief description of the extent of the inspection.
4—A description of the maintenance manual used.

6313. Who is authorized to conduct a 100-hour inspection of an aircraft?

1—Any mechanic working for a repair station.
2—A certificated mechanic with airframe and powerplant ratings.
3—A certificated repairman.
4—A certificated mechanic with a powerplant rating.

6314. Where would you find the operating conditions that make a 100-hour inspection mandatory?

1—FAR Part 91.
2—FAR Part 43.
3—AC 43.13-2A.
4—AC 65-19C.

6315. Large airplanes and turbine-powered multiengine airplanes operated under FAR Part 91, General Operating and Flight Rules, must be inspected

1—in accordance with an inspection program authorized under FAR Part 91, Subpart C.
2—annually in accordance with FAR Part 43.
3—in accordance with progressive inspection requirements of FAR Section 91.171 and FAR Part 43.
4—annually in accordance with FAR Part 43 and after each 100-hours' time in service.

6316. During the installation of a new landing gear actuating cylinder, an excessive quantity of hydraulic fluid is spilled on the main wheel tire. To prevent rapid deterioration of the tire, it should be

1—washed with a petroleum solvent, then dried with compressed air.
2—wiped with a dry cloth, then completely dried with compressed air.
3—wiped with a dry cloth followed by a washdown with soap and water.
4—washed with alcohol or lacquer thinner to neutralize the action of the hydraulic fluid.

6317. What would be the effect if the piston return spring broke in a brake master cylinder?

1—The brakes would become spongy.
2—The brake travel would become excessive.
3—The brakes would drag.
4—The brake linkage would seize.

6318. Spongy brakes are usually a result of

1—air in the system.
2—both internal and external leakage.
3—internal leakage.
4—external leakage.

6319. In brake service work, the term bleeding brakes is the process of

1—withdrawing air only from the system.
2—withdrawing fluid from the system for the purpose of removing air that has entered the system.
3—replacing lines which tend to leak small amounts of fluid.
4—eliminating excessive pedal travel by lengthening actuator rods and attendant linkages.

6320. To prevent a very rapid extension of an oleo (air/oil) shock strut after initial compression resulting from landing impact,

1—the packings and seals are designed to provide more friction between the moving parts during extension than during compression.
2—various types of valves or orifices are used which restrict the reverse fluid flow.
3—the metering pin gradually reduces the size of the orifice as the shock strut extends.
4—the air is forced through a restricted orifice in the reverse direction.

6321. A pilot reports the right brake on an aircraft grabs when the brake pedal is depressed in a normal manner. The probable cause is that the right

1—master cylinder piston rod is bent.
2—wheel actuating cylinder is leaking fluid on the brake lining.
3—master cylinder piston return spring is weak.
4—brake drum contains considerable brake lining dust.

6322. Aside from an external leak in the line, what will cause parking brakes to creep continually to the OFF position?

1—A clogged reservoir vent.
2—An internal leak in the master cylinder.
3—Insufficient hydraulic fluid in the reservoir.
4—Glazed brake linings.

6323. Why do most aircraft tire manufacturers recommend that the tubes in newly installed tires be first inflated, fully deflated, and then reinflated to the correct pressure?

1—To allow the tube to position itself correctly inside the tire.
2—To eliminate all the air between the tube and the inside of the tire.
3—To ensure that the wheel rim has been properly assembled.
4—To test the entire assembly for leaks.

6324. The metering pins in oleo (air/oil) shock struts serve to

1—lock the struts in the UP position.
2—lock the struts in the DOWN position.
3—retard the flow of oil as the struts are compressed.
4—meter the proper amount of air in the struts.

6325. After performing maintenance to an aircraft's landing gear retraction system which may have affected its ability to retract or extend properly, it is advisable to

1—restrict the aircraft to operation within gliding distance of the airport for at least the first two flights following the maintenance.
2—restrict the aircraft from carrying passengers until a test flight has been made.
3—retract and extend the landing gear at least four times during the first flight following the maintenance.
4—make an operational check with the aircraft placed on jacks.

6326. Why do tire and wheel manufacturers often recommend that the tires on split rim wheels be deflated before removing the wheel from the axle?

1—To relieve the strain on the wheel retaining nut and axle threads.
2—As a safety precaution in case the bolts that hold the wheel halves together have been damaged or weakened.
3—So that the person removing the wheel will be sure to remember to jack the wheel.
4—To remove the static load imposed upon the wheel bearings by the inflated tire.

6327. The braking action of single-disk brakes is accomplished by compressing a rotating brake disk between two opposite brake linings. How is equal pressure on both sides of the rotating disk assured?

1—By hydraulically interconnecting the brake linings located on opposite sides of the rotating disk.
2—By keeping the brake clearances closely adjusted.
3—By keying the disk loosely in the wheel so that it can move from side to side.
4—By letting the brake linings automatically equalize because of the greater rate of wear on the lining with the most pressure.

6328. If it is determined that spongy brake action is not caused by air in the brake system, what is the next most likely cause?

1—Worn brake lining.
2—Internal leakage in the master cylinder.
3—Deteriorated flexible hoses.
4—Unevenly adjusted brakes.

6329. Many brake types can be adapted to operate mechanically or hydraulically. Which type is not adaptable to mechanical operation?

1—Single-disk spot type.
2—Duo-servo type.
3—Single-servo type.
4—Expander-tube type.

6330. After replacing the blocks in an expander-tube type brake assembly, it will be necessary to

1—ensure that the radial clearance is not less than the allowable minimum.
2—revolve the star wheel adjustment until a radial clearance of 1/16 inch has been obtained between the blocks and the brake drum.
3—make sure the movement of the actuating piston presses the brake blocks against the disk.
4—replace the tube as well as the blocks.

6331. A stripe or mark extending from the rim of a wheel onto the tire

1—is a slippage mark.
2—is a balance mark.
3—indicates the tire is a high-pressure type.
4—indicates the wheel is a high-speed type.

6332. How can it be determined that all air has been purged from the brake system?

1—By operating a hydraulic unit and watching the system pressure gauge for smooth, full-scale deflection.
2—By the increased pedal movement for brake application.
3—By submerging the bleeder hose in hydraulic fluid and watching until all air bubbles have escaped.
4—By observing the reservoir sight gauge to determine if there is any fluid movement.

6333. Overinflated aircraft tires may cause damage to the

1—brake linings.
2—wheel hub.
3—brake drum.
4—wheel flange.

6334. Debooster valves are used in brake systems primarily to

1—ensure rapid application and release of the brakes.
2—reduce brake pressure and maintain static pressure.
3—reduce the pressure and release the brakes rapidly.
4—increase pressure and release the brakes rapidly.

6335. The repair for an out-of-tolerance toe-in condition of main landing gear wheels determined not to be the result of bent or twisted components consists of

1—repositioning the axle in the oleo trunnion.
2—changing the oleo assembly.
3—inserting, removing, or changing the location of washers or spacers at the center pivotal point of the scissor torque links.
4—placing shims or spacers behind the bearing of the out-of-tolerance wheel or wheels.

6336. To prevent rapid deterioration, aircraft tires should be stored in an area that is

1—dry and warm.
2—damp and cool.
3—dry and cool.
4—damp and warm.

6337. A raised H on the stem of an air valve core denotes that the valve core is a

1—hydraulic type.
2—hard-core type.
3—high-pressure type.
4—heavy-duty type.

6338. The primary purpose for balancing aircraft wheel assemblies is to

1—prevent heavy spots and reduce vibration.
2—distribute the aircraft weight properly.
3—reduce excessive wear and turbulence.
4—improve the braking action.

6339. Pressure should not be applied to an expander tube brake without the wheel being installed to prevent

1—accumulating air in the brake system.
2—damage to the brake assembly.
3—excessive loss of hydraulic fluid.
4—damage to the brake relief valve.

6340. On all aircraft equipped with retractable landing gear, some means must be provided to

1—retract and extend the landing gear if the normal operating mechanism fails.
2—extend the landing gear if the normal operating mechanism fails.
3—prevent extension of the landing gear at airspeeds greater than that determined structurally safe.
4—prevent the throttle from being reduced below a safe power setting while the landing gear is retracted.

6341. The type of hydraulic fluid to be used in a shock strut depends upon the

1—ambient temperature at the time of servicing.
2—amount of air pressure in the shock strut.
3—type of seals used in the shock strut.
4—type of fluid in the landing gear actuating systems.

6342. An automatic damping action occurs at the steer damper if for any reason the flow of high-pressure fluid is removed from the

1—outlet of the steer damper.
2—inlet of the steer damper.
3—replenishing check valve.
4—tension steering valve.

6343. What is the purpose of the torque links attached to the cylinder and piston of a landing gear oleo strut?

1—Limit compression stroke.
2—Absorb shock and reduce bounce.
3—Hold the strut in place.
4—Maintain correct wheel alignment.

6344. The removal, installation, and repair of landing gear tires by a private pilot on an aircraft owned and operated by him/her are considered to be

1—a minor alteration.
2—a minor repair.
3—a major repair.
4—preventive maintenance.

6345. Aircraft brakes requiring a large volume of fluid to operate the brakes generally

1—do not use brake pressure relief valves.
2—use independent master cylinder systems.
3—do not use brake system accumulators.
4—use power brake control valves.

6346. What is one effect a restricted compensator port of a master cylinder will have on a brake system?

1—Will operate normally.
2—Reverse fluid flow to reservoir will be unaffected.
3—The reservoir will be filled by reverse flow.
4—Will cause slow release of brakes.

6347. When an air/oil type of landing gear shock strut is used, the initial shock of landing is cushioned by

1—compression of the air charge.
2—the air charge being forced into the fluid and then released as the shock load is reduced.
3—the fluid being forced through a metered opening.
4—compression of the fluid.

6348. Internal leakage in a brake master cylinder unit can cause

1—slow release of brakes.
2—fading brakes.
3—a reduction in brake pedal travel.
4—low fluid in the master cylinder reservoir.

6349. A sleeve, spacer, or bumper ring is incorporated in a landing gear oleo shock strut to

1—limit the extension of the torque arm.
2—correct for wheel alignment.
3—limit the extension stroke.
4—reduce the rebound effect.

6350. The purpose of a sequence valve in a hydraulic retractable landing gear system is to ensure the operation at the proper time of the

1—main gear safety switches.
2—landing gear doors.
3—nose gear safety switches.
4—main gear downlocks.

6351. The pressure source for power brakes is

1—the main hydraulic system.
2—the power brake reservoir.
3—a master cylinder.
4—pressure being applied to the rudder pedals.

6352. Which statement is true with respect to an aircraft equipped with hydraulically operated multiple-disk type brake assemblies?

1—There are no minimum or maximum disk clearance checks required due to the use of self-compensating cylinder assemblies.
2—No emergency pressure system provisions are possible with this type of brake assembly.
3—Do not set parking brake when brakes are hot.
4—No parking brake provisions are possible for this type of brake assembly.

6353. What type of valve is used in the brake actuating line to isolate the emergency brake system from the normal power brake control valve system?

1—A bypass valve.
2—An orifice check valve.
3—A brake pressure relief valve.
4—A shuttle valve.

6354. When tightening the packing nut on an air/oil shock strut installed on an aircraft, the

1—packing gland should be replaced.
2—aircraft should be jacked.
3—fluid should be removed from the strut.
4—torque should not exceed 800 foot-pounds.

6355. Cracks in the surface of a brake drum

1—are always cause for rejection.
2—are not cause for rejection unless they extend the total width of the drum.
3—are not cause for rejection unless they involve more than one-half the drum width.
4—may not be cause for rejection if the cracks are less than 1 inch in length and do not extend to the edge of the drum.

6356. Instructions concerning the type of fluid and amount of air pressure to be put in a shock strut are found

1—on the airplane data plate.
2—in the aircraft operations limitations.
3—in AC 43.13-1A.
4—in the aircraft manufacturer's service manual.

6357. The purpose of a relief valve in a brake system is to

1—compensate for a drop in pressure.
2—reduce pressure for brake application.
3—prevent the tire from skidding.
4—compensate for thermal expansion.

6358. When should tire pressure be checked?

1—Immediately after each flight.
2—At least once a week or more often.
3—When tires are hot.
4—At 2-week intervals.

6359. If the extended longitudinal axis of the main landing gear wheel assemblies intersects aft of the aircraft, the wheels can be termed as having

1—toe-out.
2—toe-in.
3—negative camber.
4—positive caster.

6360. What is the purpose of a compensating port or valve in a brake master cylinder of an independent brake system?

1— Assists in the master cylinder piston return.
2— Prevents fluid from flowing back to the reservoir.
3— Assists in building up brake pressure.
4— Permits reservoir fluid to fill the master cylinder.

6361. An aircraft equipped with multiple-disk brakes was reported to have sluggish and jerky brake operation. What action should be taken?

1— Reduce the brake clearance.
2— Disassemble and clean the automatic adjuster.
3— Replace retractor springs.
4— Bleed the brakes.

6362. If an aircraft shock strut (air/oil type) bottoms upon initial landing contact, but functions correctly during taxi, the most probable cause is

1— low fluid.
2— low air charge.
3— a restricted metering pin orifice.
4— reversed metering pin.

6363. What is the function of a cam incorporated in a nose gear shock strut?

1— Provides an internal shimmy damper.
2— Straightens the nosewheel.
3— Provides jacking points.
4— Provides steering of aircraft during ground operation.

6364. Extension of an oleo (air/oil) shock strut is measured to determine the

1— amount of oil in the strut.
2— physical condition of the strut itself.
3— proper operating position of the strut.
4— mechanical efficiency of the strut.

6365. Debooster cylinders are used in brake systems primarily to

1— reduce brake pressure and maintain static pressure.
2— maintain a static pressure and ensure a rapid application.
3— relieve excessive fluid and ensure a positive release.
4— reduce the pressure to the brake and increase the volume of fluid flow.

6366. If a shock strut bottoms after it has been properly serviced, the

1— strut should be disassembled and the metering pin orifice plate replaced.
2— air pressure should be increased until the strut no longer bottoms.
3— fluid level should be increased until the strut no longer bottoms.
4— strut should be removed, disassembled, and inspected.

6367. A high-speed aircraft tire with a sound cord body and bead may be recapped

1— a maximum of three times.
2— only by the tire manufacturer.
3— and used only as a low-speed tire.
4— an indefinite number of times.

6368. If an airplane equipped with master cylinders and single-disk brakes has excessive brake pedal travel, but the brakes are hard and effective, the probable cause is

1— the master cylinder one-way cup is leaking.
2— worn brake linings.
3— the push rod between the brake pedal and the master cylinder is too long.
4— worn brake disk causing excessive clearance between the notches on the perimeter of the disk and the splines or keys on the wheel.

6369. The correct inflation pressure for an aircraft tire can be obtained from

1— tire pressure charts based on gross landing weight.
2— the operator's manual.
3— the information stamped on the wheel.
4— the aircraft's logbook.

6370. What should be checked when shock struts bottom during a landing?

1— Air pressure in the struts.
2— Struts for correct alignment.
3— Packing seals for binding.
4— Fluid level.

6371. How can it be determined that all air has been purged from a master cylinder brake system?

1—By operating a hydraulic unit and watching the system pressure gauge for smooth, full-scale deflection.
2—By noting whether the brake is firm or spongy.
3—By noting the amount of fluid return to the master cylinder upon brake release.
4—By observing the length of the brake pedal movement.

6372. The left brake is dragging excessively on an airplane on which no recent brake service work has been performed. The most probable cause is

1—foreign particles stuck in the master cylinder compensating port.
2—excessively worn brake linings.
3—too much play in the linkage interconnecting the pedals and the master cylinder.
4—low fluid supply in the brake system reservoir.

6373. If a brake debooster is used in a hydraulic brake system, its position in the system will be

1—at the brake actuating cylinder to provide connections for main and auxiliary brake systems.
2—between the pressure manifold of the main hydraulic system and the power brake control valve.
3—between the brake control valve and the brake actuating cylinder.
4—in the brake pressure line between the brake pedal and the brake accumulator.

6374. Shock struts utilizing natural rubber seals should not be serviced with

1—vegetable base fluid.
2—castor oil and alcohol.
3—hydraulic fluid which is blue in color.
4—mineral base fluid.

6375. One method of bleeding brakes which utilizes a power-brake control valve and a debooster is to charge the aircraft's hydraulic system and depress the brake pedal, with the bleeder valve at the brake open and the bleeder hose submerged in a fluid container, until all air has been purged from the system. An important consideration when bleeding by this method is to

1—bleed the debooster separately after the brake bleeding is complete.
2—recharge the debooster with dry air or nitrogen after bleeding has been completed.
3—allow the debooster to fill with fluid after the brake pedal has been released and before the bleeder valve is closed.
4—discontinue bleeding before the debooster reaches the end of its travel.

6376. If a tire has been subjected to a temperature high enough to melt one of the fusible plugs in the wheel, what should be done with the tire?

1—The tire should be scrapped.
2—The tire should be removed from the wheel and inspected before returning it to service.
3—The tire should be recapped before returning it to service.
4—The tire should be returned to service if an external inspection reveals no carcass or tread damage.

6377. To avoid nylon tire spotting on an aircraft that is to remain idle for a period longer than 3 days,

1—the aircraft should be moved every 60 hours.
2—the aircraft should be moved every 36 hours.
3—no action is necessary.
4—the aircraft should be blocked up so that no weight is on the tires or moved every 48 hours.

6378. The best safeguards against heat buildup in aircraft tires are

1—proper tire inflation, minimum braking, and fast taxi speeds.
2—short ground rolls, slow taxi speeds, minimum braking, and proper tire inflation.
3—minimum braking, proper tire inflation, and long ground rolls.
4—fast taxi speeds, short ground rolls, and coordinated turns.

6379. The fusible plugs installed in some aircraft wheels will

1— facilitate servicing of the wheel assembly.
2— indicate tire tread separation.
3— eliminate the need to check air pressure.
4— melt at a specified elevated temperature.

6380. What action, if any, should be taken when there is a difference of more than 5 pounds of air pressure in tires mounted as duals?

1— Ignore.
2— Lower the air pressure in the tire with the highest pressure to agree with its mate.
3— Correct the discrepancy and enter in logbook.
4— Replace the tire with the lowest pressure.

6381. How long should you wait after a flight before checking tire pressure?

1— At least 1 hour (2 hours in hot weather).
2— At least 2 hours (3 hours in hot weather).
3— At least 3 hours (4 hours in hot weather).
4— At least 4 hours (5 hours in hot weather).

6382. A newly mounted tire and/or tube should be

1— deflated and reinflated after 24 hours.
2— checked for proper inflation daily for several days.
3— deflated and reinflated after 12 hours.
4— checked for proper inflation after 48 hours.

6383. Excessive wear in the shoulder area of an aircraft tire is an indication of

1— overinflation.
2— excessive toe-in.
3— underinflation.
4— fast cornering during taxi.

6384. Excessive wear in the center of the tread of an aircraft tire is an indication of

1— underinflation.
2— excessive toe-out.
3— excessive braking during landing roll and taxi.
4— overinflation.

6385. When an empty shock strut is filled, care should be taken to extend and compress the strut completely at least two times to

1— thoroughly lubricate the piston rod.
2— show if there are any fluid leaks.
3— force out any excess fluid.
4— ensure proper packing ring seating and removal of air bubbles.

6386. In shock struts, chevron seals are used to

1— prevent air from escaping.
2— absorb bottoming effect.
3— prevent oil from escaping.
4— serve as a bearing surface.

6387. On most aircraft, the oil level of an air and oil shock strut is checked by

1— removing the oil filler plug and inserting a gauge.
2— measuring the overall length of the strut.
3— releasing the air and seeing that the oil is to the level of the filler plug.
4— checking the air pressure.

6388. A pilot reports that the brakes are dragging on an aircraft equipped with expander-tube type brakes. On investigation, it is found that the aircraft has been in storage for the past 6 months and the

1— brake linings have swelled due to age.
2— expander tubes have grown during the inoperative period.
3— brake drum has grown inward during the inoperative period.
4— brake lining return springs have lost their temper with age.

6389. A landing gear position and warning system will provide a warning in the cockpit when the throttle is

1— advanced and gear is not down and locked.
2— retarded and gear is not down and locked.
3— advanced and gear is down and locked.
4— retarded and gear is down and locked.

6390. An electric motor used to raise and lower a landing gear would most likely be a

1— shunt field series-wound motor.
2— split field shunt-wound motor.
3— continuous duty motor.
4— split field series-wound motor.

6391. When installing a chevron-type seal in an aircraft hydraulic cylinder, the open side of the seal should face

1— opposite the direction of fluid pressure.
2— up or forward when the unit is installed in a horizontal position.
3— the direction of fluid pressure.
4— down or aft when the unit is installed in a vertical position.

6392. Nose gear centering cams are used in many retractable landing gear systems. The primary purpose of the centering device is to

1— align the nosewheel prior to touchdown.
2— engage the nosewheel steering.
3— deactivate the nose gear torque links.
4— center the nosewheel before it enters the wheel well.

6393. What device in a hydraulic system with a constant-delivery pump allows circulation of the fluid when no demands are on the system?

1— Pressure relief valve.
2— Shuttle valve.
3— Pressure regulator.
4— Debooster valve.

6394. A fully-charged hydraulic accumulator provides

1— air pressure to the various hydraulic components.
2— a source for additional hydraulic power when heavy demands are placed on the system.
3— positive fluid flow to the pump inlet.
4— additional hydraulic fluid.

6395. A hydraulic system referred to as a "power pack" system will

1— develop more pressure per square inch than a normal system.
2— have an engine-driven pump for greater pressure.
3— have all hydraulic power components located in one unit.
4— have a pressurized reservoir.

6396. A hydraulic hose identified as MIL-H-8794 will have a yellow stripe running the length of the hose. This stripe

1— identifies the pressure rating of the hose.
2— is used to ensure that the hose is installed without excessive twisting.
3— identifies that the hose is for hydraulic fluid only.
4— identifies that the hose is constructed of synthetic rubber and may be suitable for a wide range of applications.

6397. An O-ring intended for use in a hydraulic system using MIL-H-5606 (mineral base) fluid will be marked with

1— a blue stripe or dot.
2— one or more white dots.
3— a yellow stripe.
4— a white and yellow stripe.

6398. What condition would most likely cause excessive fluctuation of the pressure gauge when the hydraulic pump is operating?

1— Bourdon tube in the gauge is broken.
2— Accumulator air pressure low.
3— Inadequate supply of fluid.
4— System relief valve sticking closed.

6399. A filter incorporating specially treated cellulose paper is identified as a

1— sediment trap.
2— finger strainer.
3— cuno filter.
4— micronic filter.

6400. The purpose of an orifice check valve is to

1— relieve pressure to a sensitive component.
2— restrict flow in one direction.
3— relieve pressure in one direction and prevent flow in the other direction.
4— relieve pressure in locked mechanisms.

6401. (Refer to figure 10.) The trunnion nut on an aircraft landing gear requires a torque of 320 inch-pounds. To reach the nut, a 2-inch straight adapter must be used on an 18-inch torque wrench. How many foot-pounds will be indicated on the torque wrench when the required torque of the nut is reached?

1— 24 foot-pounds.
2— 28.8 foot-pounds.
3— 22 foot-pounds.
4— 16 foot-pounds.

6402. A special bolt in a landing gear attachment requires a torque value of 440 inch-pounds. How many foot-pounds are required?

1— 44.0.
2— 40.0.
3— 44.6.
4— 36.6.

6403. To protect packing rings or seals from damage when it is necessary to install them over or inside threaded sections, the

1— threaded section should be coated with a heavy grease.
2— threaded section should be coated with a heavy mixture of white lead and grease.
3— packings should be compressed or stretched during installation to avoid contact with the threads.
4— threaded section should be covered with a stiff paper sleeve.

6404. To prevent external and internal leakage in aircraft hydraulic units, the most commonly used type of seal is the

1— cup seal.
2— O-ring seal.
3— rubber seal.
4— chevron seal.

6405. Which allows free fluid flow in one direction and no fluid flow in the other direction?

1— Check valve.
2— Selector valve.
3— Metering piston.
4— Master cylinder.

6406. Select the valve used in a hydraulic system that directs pressurized fluid to one end of an actuating cylinder and simultaneously directs return fluid to the reservoir from the other end.

1— Sequence.
2— Shuttle.
3— Check.
4— Selector.

6407. What function does the absolute pressure regulator perform in the pneumatic power system?

1— Regulates the compressor outlet air pressure to stabilize the system pressure.
2— Regulates the pneumatic system pressure to protect the moisture separator from internal explosion.
3— Regulates the compressor inlet air to prevent excessive speed variation and/or overspeeding of the compressor.
4— Regulates the compressor inlet air to provide a stabilized source of air for the compressor.

6408. (1) Relief valves are used in pneumatic systems as damage-preventing units.

(2) Check valves are used in both hydraulic and pneumatic systems.

Regarding the above statements,

1— both No. 1 and No. 2 are true.
2— only No. 2 is true.
3— neither No. 1 nor No. 2 is true.
4— only No. 1 is true.

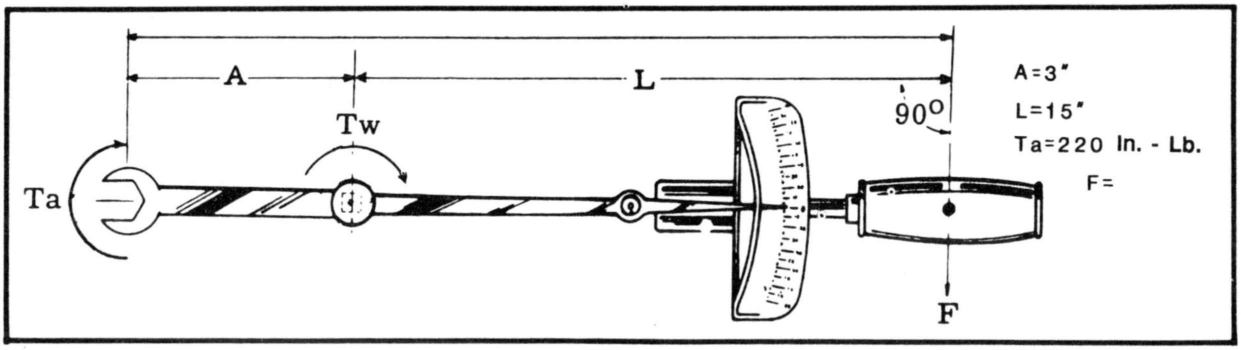

Figure 10

6409. One of the distinguishing characteristics of an open-center selector valve used in a hydraulic system is that

1—this type valve has only three ports.
2—fluid flows through the valve in the OFF position.
3—fluid flows in three directions in the ON position.
4—a limited amount of fluid flows in one direction and no fluid flows in the opposite direction.

6410. What type of packings should be used in hydraulic components to be installed in a system containing Skydrol?

1—AN packings made of natural rubber.
2—AN packings made of either natural rubber or neoprene.
3—Packing materials made for ester base fluids.
4—AN packings made of neoprene.

6411. Which should be done before removing a spool-type or a balanced-type pressure regulator from a hydraulic system?

1—Decrease the tension of the pressure regulator control spring until the unloading valve opens.
2—Reduce the accumulator air pressure to zero.
3—Move the landing gear selector valve between the DOWN and NEUTRAL positions until no pressure remains in the system.
4—Actuate the flap selector valve until no pressure remains in the system.

6412. Relief valves are used in pneumatic systems

1—for one direction flow control.
2—to control emergency airbrakes.
3—to reduce the rate of airflow.
4—as damage-preventing units.

6413. An aircraft pneumatic system, which incorporates an engine-driven multistage reciprocating compressor, also requires

1—an oil separator.
2—a surge chamber.
3—a vacuum relief valve.
4—a moisture separator.

6414. The removal of air from an aircraft hydraulic system generally presents no problem because

1—the air will bleed out if the system is allowed to remain inoperative for a period of time.
2—the air will be removed by operating the various hydraulic components through several cycles.
3—the high working pressure in the system will physically unite the air and fluid which will result in the air being dissipated in the form of heat.
4—each hydraulic actuator has a bleeder valve located on or adjacent to it for this purpose.

6415. Pneumatic systems utilize

1—hand pumps.
2—relief valves.
3—freon.
4—diluter valves.

6416. The component in the hydraulic system that is used to direct the flow of fluid is the

1—check valve.
2—orifice check valve.
3—relief valve.
4—selector valve.

6417. What type of selector valve is one of the most commonly used in hydraulic systems to provide for simultaneous flow of fluid into and out of a connected actuating unit?

1—Three-port, three-way valve.
2—Four-port, closed center valve.
3—Three-port, four-way valve.
4—Two-port, open center valve.

6418. What is the purpose of using backup rings with 0-rings in hydraulic systems above 1,500 PSI?

1—Prevent internal and external leakage of all moving parts within a hydraulic system.
2—Provide a seal between two parts of a unit which move in relation to each other.
3—Prevent high pressure from extruding the seal between the moving and stationary part.
4—Prevent loss of fluid in the event of unit failure or line breakage.

6419. The purpose of the pressure regulator in a hydraulic system is to

1— maintain system operating pressure within a predetermined range and to unload the pump.
2— maintain a constant pressure within the system and direct fluid flow to actuating cylinders.
3— regulate the amount of fluid flow to the actuating cylinders within the system.
4— prevent failure of components or rupture of hydraulic lines under excessive pressure.

6420. A flexible sealing element subject to motion is a

1— washer.
2— compound.
3— packing.
4— gasket.

6421. Which characteristics apply to aircraft hydraulic systems?

A. Minimum maintenance requirements.
B. Lightweight.
C. About 80 percent operating efficiency (20 percent loss due to fluid friction).
D. Simple to inspect.

1— A,B,C,D.
2— A,C,D.
3— A,B,D.
4— B,C,D.

6422. If a rigid tube is too short for the flare to reach its seat, you should not pull it into place by tightening the nut because

1— the threads will be stripped off of the fitting.
2— the flare may be distorted.
3— excessive stress will be imposed on the fitting.
4— vibration will work-harden the line.

6423. The installation of a new metal hydraulic line should be made with

1— a straight tube rather than one incorporating bends, for this will reduce weight and resistance to fluid flow.
2— a straight tube to withstand the shocks and vibration to which it will be subjected.
3— a straight tube to permit proper alignment of the fitting and thereby reduce fluid loss through leakage.
4— enough bends to allow the tube to expand and contract with temperature changes and to absorb vibration.

6424. Extrusion of an O-ring seal is prevented in a high-pressure system by the use of a

1— backup ring on the side of the O-ring next to the pressure.
2— U-ring on the side of the O-ring away from the pressure.
3— backup ring on the side of the O-ring away from the pressure.
4— U-ring on the side of the O-ring next to the pressure.

6425. What is one advantage of piston-type hydraulic motors over electric motors?

1— They are considerably quieter in operation.
2— They have fewer moving parts.
3— There is no fire hazard if the motor is stalled.
4— They work satisfactorily over a wider temperature range.

6426. The first step in removing an accumulator from an aircraft is to

1— relieve system pressure.
2— discharge the preload.
3— drain the reservoir.
4— purge the accumulator lines.

6427. (Refer to figure 11.) Which fitting is an AN flared-tube fitting?

1— A.
2— B.
3— C.
4— D.

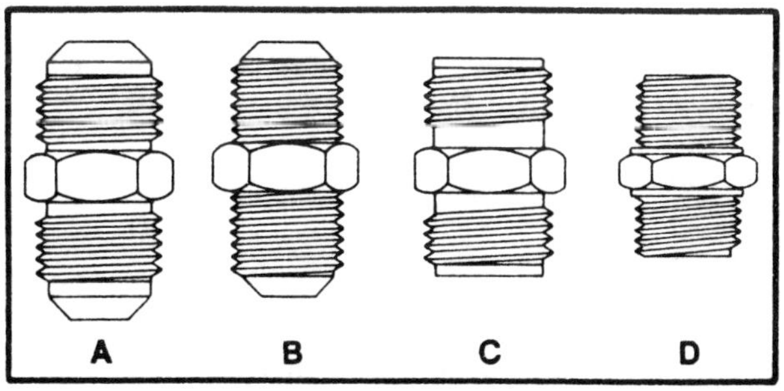

Figure 11

6428. (Refer to figure 12.) Which illustration(s) show(s) the correct spiral for teflon backup rings?

1— A.
2— B.
3— C.
4— Both A and C.

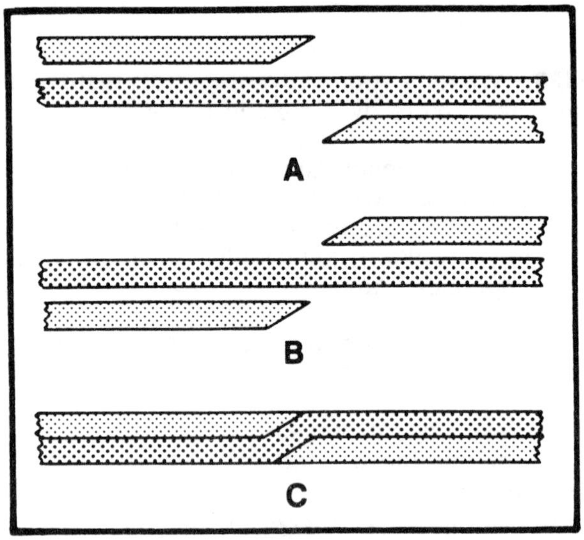

Figure 12

6429. If a hydraulic brake system uses natural rubber packing materials, the correct hydraulic fluid to service the system is

1— vegetable base oil.
2— mineral base oil.
3— synthetic oil.
4— a phosphate ester base oil.

6430. The internal resistance of a fluid which tends to prevent it from flowing is called

1— volatility.
2— flash point.
3— viscosity.
4— acidity.

6431. What is the viscosity of hydraulic fluid?

1— The increase in volume of a fluid due to temperature change.
2— The temperature at which a fluid gives off vapor in sufficient quantity to ignite momentarily.
3— The fluid's ability to resist oxidation and deterioration for long periods.
4— The internal resistance of a fluid which tends to prevent it from flowing.

6432. Which is a characteristic of petroleum base hydraulic fluid?

1— Temperature stability is very poor.
2— Flammable under normal conditions.
3— Compatible to natural rubber seals and packings.
4— Nonflammable under all conditions.

6433. (1) When servicing aircraft hydraulic systems, use the type fluid specified in the aircraft manufacturer's maintenance manual or on the instruction plate affixed to the reservoir or unit.

(2) Hydraulic fluids for aircraft are dyed a specific color for each type of fluid.

Regarding the above statements,

1— only No. 1 is true.
2— only No. 2 is true.
3— both No. 1 and No. 2 are true.
4— neither No. 1 nor No. 2 is true.

6434. Petroleum base hydraulic fluid is which color?

1— Purple.
2— Blue.
3— Colorless.
4— Red.

6435. Which has improved the properties and characteristics of petroleum base hydraulic fluid?

1— Filters.
2— Lower operating temperatures.
3— Chemical analysis.
4— Additives.

6436. Which is a characteristic of synthetic base hydraulic fluid?

1— Low moisture retention.
2— High viscosity.
3— High flash point.
4— Low flash point.

6437. Which statement about fluids is correct?

1— Any fluid will completely fill its container.
2— All fluids are considered to be highly compressible.
3— All fluids readily transmit pressure.
4— Only liquids are considered to be fluids.

6438. If an aircraft hydraulic system is serviced with the wrong fluid, the remedial procedure should be to

1—drain, flush, and never change seals.
2—drain, flush, and usually change all affected seals.
3—change seals.
4—drain and put in the correct fluid.

6439. Three types of hydraulic fluids currently being used in civil aircraft are

1—mineral base, mixed mineral and vegetable base, and vegetable base.
2—mineral base, vegetable base, and phosphate ester base.
3—mineral base, phosphate ester base, and mixed mineral and phosphate ester base.
4—mineral base, phosphate ester base, and alcohol base.

6440. Which lists only desirable properties of a good hydraulic fluid?

1—High viscosity, low flash point, chemical stability, high fire point.
2—High flash point, low viscosity, chemical stability, low fire point.
3—Low flash point, low fire point, high viscosity, chemical stability.
4—Low viscosity, chemical stability, high flash point, high fire point.

6441. Characteristics of MIL-H-8446 (Skydrol 500 A & B) hydraulic fluid are

1—blue color, vegetable base, will burn, natural rubber seals.
2—blue color, phosphate ester base, fire resistant, butyl rubber seals.
3—light purple color, phosphate ester base, fire resistant, butyl rubber seals.
4—light green color, phosphate ester base, fire resistant, butyl rubber seals.

6442. Where can information be obtained about the compatibility of fire-resistant hydraulic fluid with aircraft materials?

1—Manufacturer's technical bulletins.
2—AC 43.13-2A.
3—Aircraft manufacturer's specifications.
4—AC 43.13-1A.

6443. Characteristics of MIL-H-5606 hydraulic fluid are

1—light purple color, phosphate ester base, fire resistant, butyl rubber seals.
2—blue color, phosphate ester base, fire resistant, butyl rubber seals.
3—blue color, vegetable base, will burn, natural rubber seals.
4—red color, petroleum base, will burn, synthetic rubber seals.

6444. Characteristics of MIL-H-7644 hydraulic fluid are

1—red color, petroleum base, will burn, synthetic rubber seals.
2—light purple color, phosphate ester base, fire resistant, butyl rubber seals.
3—blue color, vegetable base, will burn, natural rubber seals.
4—blue color, phosphate ester base, fire resistant, butyl rubber seals.

6445. If an aircraft hydraulic system requires vegetable base hydraulic fluid, but phosphate ester base hydraulic fluid is used, what will be the effect on the system?

1—No effect.
2—MIL-H-8446 may be blended with MIL-H-7644 up to a 50-50 basis with no adverse effects.
3—System will be contaminated, fluids will not blend, and the seals will fail.
4—System will be contaminated, fluids will not blend, but there will be no seal problem.

6446. What is used to flush a system normally serviced with MIL-H-5606 hydraulic fluid?

1—Alcohol or toluene.
2—Methyl ethyl ketone or kerosene.
3—Naphtha or varsol.
4—Lacquer thinner or trichlorethylene.

6447. What is used to flush a system normally serviced with Skydrol hydraulic fluid?

1—Trichlorethylene.
2—Alcohol.
3—Naphtha.
4—Kerosene.

6448. How can the proper hydraulic fluid to be used in an airplane be determined?

1—Check the color code on the system lines.
2—Refer to the aircraft parts manual.
3—Consult the aircraft Type Certificate Data Sheet.
4—Consult the aircraft service manual.

6449. Phosphate ester base hydraulic fluid is very susceptible to contamination from

1— teflon seal material.
2— water in the atmosphere.
3— ethylene-propylene elastomers.
4— carbon dioxide.

6450. (1) Materials which are Skydrol compatible or resistant include most common aircraft metals and polyurethane and expoy paints.

(2) Skydrol hydraulic fluid is compatible with natural fibers and also with nylon and polyester.

Regarding the above statements,

1— neither No. 1 nor No. 2 is true.
2— both No. 1 and No. 2 are true.
3— only No. 1 is true.
4— only No. 2 is true.

6451. The hydraulic component that automatically directs fluid from either the normal source or an emergency source to an actuating cylinder is called a

1— time-lag valve.
2— bypass valve.
3— shuttle valve.
4— crossflow valve.

6452. What is the primary purpose of a hydraulic actuator?

1— Transmit hydraulic fluid.
2— Transform fluid pressure into mechanical force.
3— Transform fluid pressure into static force.
4— Transfer fluid from the pressure side to the return side of a hydraulic system.

6453. The primary function of the flap overload valve is to

1— prevent the flaps from being lowered at airspeeds which would impose excessive structural loads.
2— ensure that the flaps will not change their angular setting due to internal leakage of the flap selector valve.
3— cause the flap segments located on opposite sides of the aircraft centerline to extend and retract together so that the aircraft will not become aerodynamically unbalanced to the extent that it becomes uncontrollable.
4— boost normal system pressure to the flaps in order to overcome the air loads acting on the relatively large flap area.

6454. A unit which transforms hydraulic pressure into linear motion is called

1— an actuating cylinder.
2— an accumulator.
3— a hydraulic pump.
4— a master cylinder.

6455. If it is necessary to adjust several pressure regulating valves in a hydraulic system, what particular sequence, if any, should be followed?

1— Units most distant from the hydraulic pump should be adjusted first.
2— Units with the highest pressure settings are adjusted first.
3— Units are independent of each other, and therefore, no particular sequence is necessary.
4— Units are adjusted in order of the lowest specified pressure settings.

6456. How are the open ends of hydraulic lines protected from contamination after a hydraulic unit has been removed?

1— Seal the lines with caps or plugs.
2— Seal the lines by inserting a piece of cloth with a red tag attached into the open end.
3— Seal the lines with masking tape.
4— Remove the section of line on each side of the valve.

6457. If an aircraft's constant-pressure hydraulic system cycles more frequently than usual and no unusual fluid leakage can be detected, the most probable cause is

1— excessively high relief valve setting.
2— moisture absorbed in the fluid.
3— low accumulator air preload.
4— an obstruction in the reservoir vent line.

6458. Unloading valves are used with many engine-driven hydraulic pumps to

1— prevent excessive loss of fluid.
2— dampen out pressure surges.
3— relieve the pump pressure.
4— relieve system pressure.

6459. What is the purpose of the shear section on the shaft of an engine-driven hydraulic power pump?

1— It permits easy alignment of the splined drive shaft with the engine accessory drive splines.
2— It absorbs shock loads caused by rapid changes in engine RPM or hydraulic system pressure rises.
3— It prevents pressure surges in the hydraulic system from overloading the pump components.
4— It allows the shaft to break if the pump should seize.

6460. Which valve installed in a hydraulic system will have the highest pressure setting?

1— Pressure regulator valve.
2— Main relief valve.
3— Thermal relief valve.
4— Pump unloading valve.

6461. Excluding lines, which components are required to make up a simple hydraulic system?

1— Actuator, pressure reservoir, accumulator, and selector valve.
2— Pump, reservoir, selector valve, and actuator.
3— Pump, reservoir, relief valve, and shuttle valve.
4— Hydraulic motor, selector actuator, and pressure gauge.

6462. Most variable displacement hydraulic pumps of current design

1— must be driven at a nearly constant speed in order to be practical for use.
2— are not practical for use with a closed-center hydraulic system.
3— contain a built-in means of system pressure regulation.
4— cannot be used except in multiengine installations due to their unreliability.

6463. In a gear-type hydraulic pump, a mechanical safety device incorporated to protect the pump from overload is the

1— thermal expansion valve.
2— bypass valve.
3— check valve.
4— shear pin.

6464. After installation of a rebuilt hydraulic hand pump, it is found that the handle cannot be moved in the pumping direction (pressure stroke). The most likely cause is an incorrectly installed

1— hand pump inport check valve.
2— piston cup seal.
3— piston rod displacement valve.
4— hand pump outport check valve.

6465. Pressure is a term used to indicate the force per unit area. Pressure is usually expressed in

1— pounds per square inch.
2— pounds per inch.
3— pounds per cubic inch.
4— pounds.

6466. If two actuating cylinders which have the same cross-sectional area but different lengths of stroke are connected to the same source of hydraulic pressure, they will exert

1— different amounts of force but will move at the same rate of speed.
2— equal amounts of force but will move at different rates of speed.
3— equal amounts of force and will move at the same rate of speed.
4— different amounts of force and will move at different rates of speed.

6467. Using a hand pump, pressure of 100 PSI has been built up in a hydraulic system. The hand pump piston is 1 inch in diameter. A 1/2-inch line connects the hand pump to an actuating cylinder 2 inches in diameter. What is the pressure in the line between the hand pump and the actuator?

1— 50 PSI.
2— 100 PSI.
3— 150 PSI.
4— 200 PSI.

6468. The spool-type hydraulic system pressure regulator

1—will retain pressure in the system for a predetermined period of time after engagement by the pilot or other crewmember.
2—is automatic in operation and normally requires no attention by the pilot or flightcrew.
3—must be separately engaged after the selector valve for the component to be actuated has been moved to the desired position.
4—must be separately engaged before actuation of any hydraulically operated component.

6469. Which is true regarding the ground check of a flap operating mechanism which has just been installed?

1—Binding of the mechanism will be indicated by the slow, steady travel of the flap from one position to another.
2—If the time required to operate the mechanism increases with successive operations, it indicates the air is being worked out of the system.
3—If the time required to operate the mechanism decreases with successive operations, it indicates the air is being worked out of the system.
4—All hydraulic lines and components should be checked for leaks by applying soapy water to all connections.

6470. A hydraulic system operational check during ground runup of an aircraft indicates that the wing flaps cannot be lowered using the main hydraulic system, but can be lowered by using the emergency hand pump. Which is the most likely cause?

1—The flap selector valve has a severe internal leak.
2—The pressure accumulator is not supplying pressure to the system.
3—The fluid level in the reservoir is low.
4—The main system relief valve setting is too low.

6471. Many hydraulic reservoirs contain a small quantity of fluid which is not available to the main system pump. This fluid is retained to

1—prime the main system.
2—supply fluid to the auxiliary pump.
3—maintain a positive head of pressure.
4—supply fluid to the pressure accumulator.

6472. The unit which causes one hydraulic operation to follow another in a definite order is called a

1—time-lag valve.
2—sequence valve.
3—follower valve.
4—crossflow valve.

6473. The purpose of a hydraulic pressure regulator is to

1—prevent the system pressure from rising above a predetermined amount due to thermal expansion.
2—direct the entire pump output to the essential units in case of emergency.
3—boost the pressure in the portions of the system that require higher pressure.
4—relieve the pump of its load when no actuating units are being operated.

6474. Severe kickback of the emergency hydraulic hand pump handle during the normal intake stroke will indicate which of the following?

1—The hand pump inport check valve is sticking open.
2—The main system relief valve is set too high.
3—The accumulator still contains its normal air charge and the emergency hand pump cannot override it.
4—The hand pump outport check valve is sticking open.

6475. What type of valve in an aircraft hydraulic system permits fluid to flow freely in one direction, but restricts the rate at which fluid is allowed to flow in the other direction?

1—Shuttle valve.
2—Check valve.
3—Orifice restrictor.
4—Orifice check valve.

6476. The main system pressure relief valve in a simple hydraulic system equipped with a power control valve should be adjusted

1—after the power control valve automatic kick-out pressure has been set.
2—with the power control valve held in the CLOSED position.
3—while one or more actuating units are in operation.
4—with the power control valve in the OPEN position.

6477. A hydraulic accumulator is charged with an air preload of 1,000 PSI. W' ən a hydraulic system pressure of 3,000 PSI is developed, the pressure on the air side of the accumulator will be

1— 1,000 PSI.
2— 2,000 PSI.
3— 3,000 PSI.
4— 4,000 PSI.

6478. How is the air in a hydraulic accumulator prevented from entering the fluid system?

1— By forcing the oil/air mixture through a centrifugal separating chamber that prevents the air from leaving the accumulator.
2— By mounting the accumulator with the oil outlet at the bottom so that it will always be covered by the oil.
3— By physically separating the air chamber from the oil chamber with a flexible or movable separator.
4— By including a float-operated valve that automatically closes when the fluid level lowers to a preset amount.

6479. After a hydraulic accumulator has been installed and air chamber charged, the main system hydraulic pressure gauge will not show a hydraulic pressure reading until

1— at least one selector valve has been actuated to allow fluid to flow into the fluid side of the accumulator.
2— the check valve located between the accumulator and the pressure manifold has been unseated.
3— the air pressure has become equal to the fluid pressure.
4— the fluid side of the accumulator has been charged.

6480. Which must be done before adjusting the relief valve of a main hydraulic system incorporating a pressure regulator?

1— Eliminate the action of the unloading valve.
2— Adjust all other system relief valves which have a lower pressure setting.
3— Manually unseat all system check valves to allow unrestricted flow in both directions.
4— Remove all the system fluid, flush the system, and refill with clean fluid.

6481. Which seals are used with vegetable base hydraulic fluids?

1— Synthetic rubber.
2— Butyl rubber.
3— Neoprene rubber.
4— Natural rubber.

6482. One advantage claimed for pneumatic power systems is that no return lines are required. What happens to the air that is expended and no longer needed when an actuating unit is operated?

1— It is exhausted or dumped, usually overboard.
2— It is stored in a common low-pressure container to be reclaimed during the next regular servicing period.
3— It is stored in lightweight bottles or containers located near each actuator.
4— It is charged or pressurized for use during the next operating cycle.

6483. Some hydraulic systems incorporate a device which is designed to remain open to allow a normal fluid flow in the line, but closed if the fluid flow increases above an established rate. This device is generally referred to as a

1— shuttle valve.
2— hydraulic fuse.
3— flow regulator.
4— metering check valve.

6484. When hydraulic system pressure control and relief units fail to function properly, how are most systems protected against overpressure?

1— A shear section on the main hydraulic pump drive shaft.
2— One or more hydraulic fuses installed in the pressure and return lines.
3— Controlled leakage past the main hydraulic pump drive shaft packings.
4— A shuttle valve interconnecting the main and emergency systems.

6485. A worn hydraulic pump shaft seal can normally be detected by

1— hydraulic fluid flowing from the pump drain line.
2— evidence of engine oil combined in the hydraulic fluid.
3— evidence of hydraulic fluid combined in the engine oil.
4— the presence of hydraulic fluid around the pump mounting pad.

6486. If an engine-driven hydraulic pump of the correct capacity fails to maintain normal system pressure during the operation of a cowl flap actuating unit, the probable cause is

1— mechanical interference to the movement of the cowl flap.
2— severe bends in the cowl flap actuating cylinder lines.
3— a partial restriction in the inport of the selector valve.
4— restriction in the pump outlet.

6487. Before removing the filler cap of a pressurized hydraulic reservoir,

1— relieve the hydraulic system pressure.
2— actuate several components in the system.
3— relieve the air pressure.
4— disconnect all electrical power.

6488. What happens to the output of a constant-displacement hydraulic pump when the hydraulic system pressure regulator diverts the fluid from the system to the reservoir?

1— The output pressure remains the same, but the volume reduces.
2— The output pressure and volume reduce.
3— The output pressure reduces, but the volume remains the same.
4— The output pressure and volume remain the same.

6489. The primary function of the hydraulic system accumulator is to

1— act as an emergency supply of fluid for the hand pump.
2— ensure a continuous supply of fluid to the engine-driven pump.
3— relieve the pump of the constant load when no units in the system are operating.
4— supplement the pump when it is under peak load.

6490. Chattering of the hydraulic pump during engine runup is an indication

1— that the pressure gauge snubber is inoperative.
2— that the main system relief valve is sticking open.
3— that the ball-check valve has not been installed at the pump outlet.
4— of an air leak in the pump inlet line.

6491. At which of the following places should you expect to find hydraulic quick-disconnect fittings?

1— In the wheel wells.
2— At the hydraulic reservoir.
3— At the hydraulic pumps.
4— On the firewall.

6492. Which seal/material is used with phosphate ester base hydraulic fluids?

1— Natural rubber.
2— Synthetic rubber.
3— Butyl rubber.
4— Neoprene rubber.

6493. A hydraulic pump is a constant-displacement type if it is able to

1— supply the demands of the system.
2— produce an unregulated constant pressure.
3— produce a continuous positive pressure.
4— deliver a uniform rate of fluid flow.

6494. A hydraulic motor converts fluid pressure to

1— linear motion.
2— rotary motion.
3— angular motion.
4— vertical motion.

6495. A crossflow valve which is designed to bypass fluid from one side of an actuating cylinder to the other side, under certain conditions, may be found in some aircraft installed in the

1— engine cowl flap system.
2— landing gear system.
3— flap overload system.
4— brake system.

6496. Hydraulic fluid filtering elements constructed of porous paper are normally

1— cleaned by turning the T-shaped handle which protrudes from the top of the filter housing.
2— discarded at regular intervals and replaced with new filtering elements rather than being cleaned.
3— not approved for use in certificated aircraft.
4— cleaned automatically by the action of the check valves located in the filter housing which permits a flow of fluid in the reverse direction each time the pressure falls below a predetermined value.

6497. A pilot reports that when the hydraulic pump is running, the pressure is normal. However, when the pump is stopped, no hydraulic pressure is available. This is an indication of a

1— leaking selector valve.
2— low accumulator fluid preload.
3— restricted pressure line.
4— leaking accumulator air valve.

6498. If fluid is added to a nonpressurized reservoir in a constant-pressure h draulic system while the system is pressurized,

1— fluid will spray violently out of the reservoir when the filler neck cap is removed.
2— the engine pump will lose its prime and will cavitate.
3— the fluid level will increase when system pressure is reduced.
4— air will be drawn into the system, when the filler neck cap is removed, resulting in an air lock in the system.

6499. In a hydraulic system that has a reservoir pressurized with turbine-engine compressor bleed air, which unit reduces the air pressure between the engine and reservoir?

1— Bellowfram bypass valve.
2— Relief valve.
3— Air bleed relief valve.
4— Air pressure regulator.

6500. What is the main purpose of a pressurized reservoir in a hydraulic system?

1— Prevent tank collapse.
2— Prevent hydraulic pump cavitation.
3— Prevent hydraulic fluid from foaming.
4— Prevent normal accumulator pressure from slaving.

6501. Two hydraulic actuating cylinders which have the same length of stroke but different diameters are connected to the same source. The two cylinders will operate at

1— the same internal pressure but will exert different amounts of force.
2— different internal pressures and will exert different amounts of force.
3— the same internal pressure and will exert equal force.
4— different internal pressures but will exert equal force.

6502. Hydraulic fluid reservoirs are sometimes designed with a standpipe in one of the outlet ports in order to assure emergency supply of fluid. The outlet port with the standpipe in it furnishes fluid to the

1— emergency pump or the normal system pump, whichever is selected.
2— emergency pump when the fluid supply to the normal system has been depleted.
3— emergency pump at any time it is required.
4— normal system power pump.

6503. An emergency supply of fluid is often retained in the main hydraulic system reservoir by the use of a standpipe located in the

1— reservoir filler neck extension.
2— inlet from the main hydraulic system.
3— outlet to the emergency pump.
4— outlet to the main system pump.

6504. To check the air charge in a hydraulic accumulator,

1— reduce all hydraulic pressure, then observe the reading on the accumulator air gauge.
2— observe the first reading on the hydraulic system gauge while operating a component in the system.
3— build up hydraulic pressure and use an air gauge on the accumulator air fitting.
4— read it directly from the auxiliary pressure gauge.

6505. How would the air pressure charge in the accumulator be determined if the engine is inoperative, but the system still has hydraulic pressure?

1— Read it directly from the main system pressure gauge with all actuators inoperative.
2— Build up system pressure with the emergency pump and then read the pressure on a gauge attached to the air side of the accumulator.
3— Operate some hydraulic unit slowly and note the pressure at which a rapid pressure drop begins as it goes toward zero.
4— Build up system pressure with the emergency pump. The system pressure gauge will remain at zero until the pressure equals accumulator air pressure, at which time the gauge reading will increase rapidly.

6506. How is phosphate ester base hydraulic fluid removed from aircraft tires?

1— Wash with soap and water.
2— Wash with petroleum solvent.
3— Wash with acetone.
4— Wash with alcohol.

6507. How many of these seals are used with petroleum base hydraulic fluids?

A. Synthetic rubber.
B. Butyl rubber.
C. Natural rubber.
D. Neoprene rubber.

1— One.
2— Two.
3— Three.
4— Four.

6508. What is used to remove petroleum base hydraulic fluid from aircraft tires?

1— Acetone.
2— Soap and water.
3— Petroleum solvent.
4— Alcohol.

6509. Hydraulic system thermal relief valves are set to open at

1— a lower pressure than the system relief valve.
2— a higher pressure than the system relief valve.
3— a lower pressure than the system pressure regulator.
4— the same pressure required to open the system relief valve.

6510. Chatter in a hydraulic system is caused by

1— excessive system pressure.
2— insufficient system pressure.
3— air in the system.
4— improper adjustment of the pressure regulator.

6511. If hydraulic fluid is released when the air valve core of the accumulator is depressed, it is evidence of

1— excessive accumulator air pressure.
2— excessive system pressure.
3— a leaking check valve.
4— a ruptured diaphragm.

6512. Although dents in the heel of a bend are not permissible, they are acceptable in the remainder of the hydraulic tube providing they do not exceed

1— 5 percent of the tube diameter.
2— 10 percent of the tube diameter.
3— 20 percent of the tube diameter.
4— 25 percent of the tube diameter.

6513. If the hydraulic system pressure is normal while the engine-driven pump is running, but there is no pressure after the engine has been shut off, it indicates

1— the system relief valve setting is too high.
2— no air pressure in the accumulator.
3— the pressure regulator is set too high.
4— the presence of air in the system.

6514. The purpose of restrictors in hydraulic systems is to

1— control the rate of movement of hydraulically operated mechanisms.
2— allow the free flow of fluid in one direction only.
3— restrict the travel range of hydraulically operated mechanisms.
4— lower the operating pressure of selected components.

6515. A common cause of slow actuation of hydraulic components is

1— cold fluid.
2— congealed fluid.
3— a weak hydraulic pump.
4— internal leakage in the actuating unit.

6516. A loud hammering noise in a hydraulic system having an accumulator usually indicates

1— air in the fluid.
2— too much preload in the accumulator.
3— relief valve set too high.
4— too low or no preload in the accumulator.

6517. Teflon hose that has developed a permanent "set" from being exposed to high pressure or temperature

1— should not be straightened or bent further.
2— should not be reinstalled once removed.
3— should be immediately replaced.
4— requires neither special handling precautions nor replacement.

6518. In a typical high-pressure pneumatic system, if the moisture separator does not vent accumulated water when the compressor shuts down, a likely cause is

1— a saturated chemical dryer.
2— a malfunctioning pressure transmitter.
3— a frozen safety fitting.
4— a malfunctioning solenoid dump valve.

6519. Which section of a turbojet engine provides air for the pressurization and air-conditioning systems of a jet aircraft?

1— Exhaust.
2— Compressor.
3— Combustion.
4— Intake.

6520. Which component of an air-cycle cooling system undergoes a pressure and temperature drop of air during operation?

1—Water separator.
2—Expansion turbine.
3—Primary heat exchanger.
4—Refrigeration bypass valve.

6521. In a freon vapor-cycle cooling system, where is cooling air obtained for the condenser?

1—Turbine engine compressor.
2—Ambient air.
3—Subcooler air.
4—Pressurized cabin air.

6522. What is ventilating air used for on a combustion heater?

1—Provides combustion air for ground blower.
2—Carries heat to the places where needed.
3—Keeps overheat thermoswitch cool.
4—Provides oxygen required to support the flame.

6523. Turbine engine air used for air-conditioning and pressurization is generally called

1—compressed air.
2—ram air.
3—conditioned air.
4—bleed air.

6524. When the cabin pressure regulator is operating in the differential mode, reference pressure is vented to the atmosphere by the

1—dump valve.
2—relief valve.
3—isobaric metering valve.
4—differential metering valve.

6525. In the combustion heater, combustion air system, what prevents too much air from entering the heaters as air pressure increases?

1—Either a combustion air relief valve or a differential pressure regulator.
2—Only a differential pressure regulator can be used.
3—Only a combustion air relief valve can be used.
4—Both a combustion air relief valve and a differential pressure regulator in every system.

6526. When the cabin pressure regulator is operating in the isobaric range, cabin pressure is maintained constant by

1—the action of the solenoid air valve.
2—the action of the cabin pressure safety valve.
3—regulating the flow of air into the cockpit and cabin.
4—the movement of the regulator bellows, limiting the flow of air from the reference chamber.

6527. What controls the operation of the cabin pressure regulator?

1—Cabin air pressure.
2—Ram air pressure.
3—Bleed air pressure.
4—Compression air pressure.

6528. The basic air-cycle cooling system consists of

1—a source of compressed air, heat exchangers, and a turbine.
2—heaters, coolers, and compressors.
3—ram air source, compressors, and engine bleeds.
4—heat exchangers and evaporators.

6529. The purpose of the dump valve in a pressurized aircraft is to relieve

1—all positive pressure from the cabin.
2—a negative pressure differential.
3—the load on the compressors.
4—pressure in excess of the maximum differential.

6530. What component might possibly be damaged if liquid refrigerant is introduced into the low side of a vapor-cycle cooling system when the pressure is too high or the outside air temperature is too low?

1—Compressor.
2—Receiver-dryer.
3—Condenser.
4—Evaporator.

6531. How can it be determined that a vapor-cycle cooling system is charged with the proper amount of freon?

1—Air bubbles in the sight glass disappear.
2—The compressor loads up and RPM decreases.
3—The evaporator has a slight frost on its exterior.
4—The compressor RPM increases.

6532. When charging a vapor-cycle cooling system after evacuation, the low-pressure gauge fails to come out of a vacuum. What is indicated?

1—Blockage in the system.
2—Normal indication.
3—The expansion valve failed to close.
4—The compressor is not engaging.

6533. What component in a vapor-cycle cooling system would most likely be at fault if a system would not take a freon charge?

1—Evaporator.
2—Expansion valve.
3—Condenser.
4—Receiver-dryer.

6534. Frost or ice buildup on a vapor-cycle cooling system evaporator would most likely be caused by

1—too much freon in the system.
2—moisture in the evaporator.
3—inadequate airflow through the evaporator.
4—high humidity.

6535. What test is used to determine the serviceability of an oxygen cylinder?

1—Pressure test with high-pressure nitrogen.
2—Pressure test with compressed air.
3—Pressure test with oxygen.
4—Pressure test with water.

6536. How often should standard weight high-pressure oxygen cylinders be hydrostatically tested?

1—Every 5 years.
2—Every 4 years.
3—Every 3 years.
4—Every 12 years.

6537. To be eligible for recharging, a DOT 3HT oxygen cylinder must have been hydrostatically tested within

1—3 years.
2—5 years.
3—7 years.
4—10 years.

6538. What type oxygen system uses the rebreather bag principle?

1—Pressure demand.
2—Diluter demand.
3—Continuous flow.
4—Demand.

6539. The positive-displacement type compressor that is most suitable for use in large piston-engined aircraft for cabin pressurization is the

1—centrifugal turbocompressor.
2—reciprocating type.
3—vane type.
4—roots blower.

6540. In large aircraft that use positive-displacement compressors for cabin pressurization, how is oil-free delivery of air achieved?

1—By using separate bearing chambers, rubber seals, and labyrinth seals in the compressor.
2—By the use of air-oil separators.
3—By using dry-type (no lubricant) pumps.
4—By using a special dry lubricant in the compressor.

6541. For emergency or backup use in pressurized aircraft, which is generally the least complicated and requires the least maintenance?

1—Liquid oxygen systems.
2—Chemical oxygen candle systems.
3—High-pressure oxygen systems.
4—Low-pressure oxygen systems.

6542. The main cause of contamination in gaseous oxygen systems is

1—moisture.
2—dirt.
3—fungus growth.
4—ozone.

6543. Where does the last stage of cooling in an air-cycle air-conditioning system occur?

1—The refrigeration unit compressor.
2—The secondary heat exchanger.
3—The expansion turbine.
4—The temperature controller.

6544. The point at which freon flowing through a vapor-cycle cooling system gives up heat and changes from a gas to a liquid is the

1—condenser.
2—evaporator.
3—compressor.
4—expansion valve.

6545. The point at which freon flowing through a vapor-cycle cooling system absorbs heat and changes from a liquid to a gas is the

1—condenser.
2—evaporator.
3—compressor.
4—expansion valve.

6546. How is the cabin pressure of a pressurized aircraft usually controlled?

1—By a valve that stops the pressurization pump when a pressure equivalent to the maximum safe cabin altitude has been reached.
2—By a pressure-sensitive switch that causes the pressurization pump to turn on or off as required.
3—By an automatic outflow valve that dumps all the pressure in excess of the amount for which it is set.
4—By a pressure-sensitive valve that controls the output pressure of the pressurization pump.

6547. Which is considered a good practice concerning the inspection of heating and exhaust systems of aircraft utilizing a jacket around the engine exhaust as a heat source?

1—All exhaust and heating system components should be replaced at each engine overhaul period.
2—Supplement physical inspections with periodic operational carbon monoxide detection tests.
3—All exhaust system components should be removed periodically, and their condition determined by the magnetic-particle inspection method.
4—All exhaust system components should be removed and replaced at each 100-hour inspection period.

6548. What will result if auxiliary (ambient) ventilation is selected during pressurized flight while at cruising altitude?

1—An increase in cabin pressure.
2—Cabin compressor overspeed.
3—Increased cabin altitude.
4—Increased conditioned air efficiency.

6549. The cabin pressure control setting has a direct influence upon the

1—outflow valve opening.
2—cabin supercharger compression ratio.
3—pneumatic system pressure.
4—turbocompressor speed.

6550. The function of the evaporator in a freon cooling system is to

1—liquefy freon in the line between the compressor and the condenser.
2—lower the temperature of the cabin air.
3—transfer heat from the freon gas to ambient air.
4—evaporate water and other impurities from the liquid freon.

6551. What is the purpose of a mixing valve in an air-conditioning system?

1—Dehumidify cabin air by mixing it with dry air.
2—Control the supply of hot, cool, and cold air.
3—Distribute conditioned air evenly to all parts of the cabin.
4—Combine emergency ram air with conditioned air.

6552. What component of a pressurization system prevents the cabin altitude from becoming higher than airplane altitude?

1—Cabin rate-of-descent control.
2—Negative pressure relief valve.
3—Supercharger overspeed valve.
4—Compression ratio limit switch.

6553. If the liquid level gauge in a vapor-cycle cooling system indicates a low freon charge, the system should

1—be operated and a pressure check performed.
2—be operated for a period of time to reach a stable condition and then the freon level rechecked.
3—not be operated until freon and oil have been added.
4—be operated and the expansion valve adjusted to permit a greater freon flow into the condenser.

6554. If the cabin rate of climb is too great, the controls should be adjusted to cause the

1—outflow valve to close slower.
2—cabin compressor speed to increase.
3—outflow valve to close faster.
4—cabin compressor speed to decrease.

6555. The position of the thermostatic expansion valve in a vapor-cycle cooling system is determined by temperature and pressure of the

1— freon entering the evaporator.
2— air in the outlet of the condenser.
3— air in the outlet of the thermostatic expansion valve.
4— freon in the outlet of the evaporator.

6556. The function of the condenser in a freon cooling system is to

1— transfer heat from the freon gas to ambient air.
2— remove water vapor from the cabin air to prevent icing of the evaporator.
3— change liquid freon into a gas before it enters the compressor.
4— transfer heat from the cabin air to the liquid freon.

6557. The function of an expansion valve in a freon cooling system is to act as a metering device and to

1— reduce the pressure of the gaseous freon.
2— increase the pressure of the liquid freon.
3— increase the pressure of the gaseous freon.
4— reduce the pressure of the liquid freon.

6558. Which prevents a loss of pressurization through a disengaged cabin air compressor?

1— Firewall shutoff valve.
2— Supercharger disconnect mechanism.
3— Cabin pressure outflow valve.
4— Delivery air duct check valve.

6559. When servicing an air-conditioning system that has lost all of its freon, it is necessary to

1— check oil and add as necessary, evacuate the system, relieve vacuum, and add freon.
2— check oil and add as necessary, evacuate the system, and add freon.
3— check oil and add as necessary, and add freon.
4— evacuate the system, and add freon.

6560. The primary function of the cabin pressurization system outflow valve is to

1— provide constant volume cabin air outflow.
2— provide protection against overpressurization.
3— maintain the desired cabin pressure.
4— regulate cabin ram air pressure.

6561. The purpose of a jet pump in a pressurization and air-conditioning system is to

1— produce a high pressure for operation of the outflow valve.
2— provide for augmentation of airflow in some areas of the aircraft.
3— assist in the circulation of freon.
4— control the output of the positive displacement cabin compressor.

6562. After cleaning or replacing the filtering element in a combustion heater fuel system, the system should be pressurized and

1— the fuel flow control valve adjusted to account for the lowered resistance to fuel flow offered by the new filtering element.
2— all connections checked for leaks.
3— the fuel filter bypass valve reset to the filter position.
4— a sample of fuel taken downstream from the filter to ensure proper operation of the new filtering element.

6563. The operation of an aircraft combustion heater is usually controlled by a thermostat circuit which

1— alternately turns the fuel on and off, a process known as cycling.
2— meters the amount of fuel continuously entering the heater and therefore regulates the heater's BTU output.
3— regulates the voltage applied to the heater's ignition transformer.
4— operates a duct damper and therefore allows only the required part of the heater's output to circulate in the cabin ducts; the remainder is dumped overboard.

6564. The air-cycle cooling system produces cold air by

1— routing conditioned air through the cooling fan.
2— passing heated air through a compressor.
3— passing air through cooling coils that contain a refrigerant.
4— extracting heat energy across an expansion turbine.

6565. (Refer to figure 13.) Determine what unit is located immediately downstream of the expansion valve in a freon refrigeration system.

1— Condenser.
2— Compressor.
3— Cooling turbine.
4— Evaporator coils.

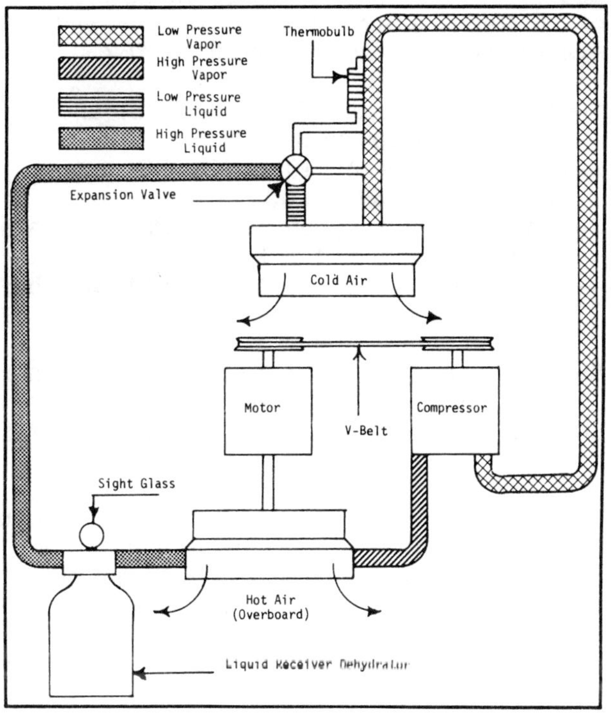

Figure 13

6566. If the isobaric metering valve in a pressurization control system moves toward the OPEN position,

1— cabin pressure will increase.
2— cabin altitude will decrease.
3— cabin pressure will decrease.
4— the outflow valve will close.

6567. When checking a freon system, a steady stream of bubbles in the sight gauge indicates

1— there is a little too much charge.
2— the charge is extremely high.
3— the proper amount of air is in the system.
4— the charge is low.

6568. Which of the following articles of safety gear must be worn when working with Freon 12?

1— Particle mask.
2— Ear plugs.
3— Rubber-soled safety shoes.
4— Goggles.

6569. An aircraft fuselage is subjected to five major stresses. Pressurization is classified as a

1— tension stress.
2— compression stress.
3— torsion stress.
4— shear stress.

6570. (1) Cabin differential pressure is the ratio between the internal and external air pressures acting on an aircraft.

(2) Cabin differential pressure is the measure of internally applied stress on the fuselage.

Regarding the above statements,

1— only No. 1 is true.
2— both No. 1 and No. 2 are true.
3— only No. 2 is true.
4— neither No. 1 nor No. 2 is true.

6571. The purpose of evacuating a freon air-conditioner system is to

1— check for leaks.
2— remove all stale, worn-out liquid.
3— assure that all check valves are sealed.
4— remove moisture.

6572. The operating cabin pressurization ranges of a pressurized aircraft are

1— isobaric, differential, and maximum differential.
2— differential, unpressurized, and isobaric.
3— ambient, unpressurized, and isobaric.
4— unpressurized, differential, and ambient.

6573. (1) Usually the air bled from a gas-turbine engine compressor is free from contamination and can be used safely for cabin pressurization.

(2) Independent cabin compressors can be engine-driven through accessory drive gearing or can be powered by bleed air from a turbine engine compressor.

Regarding the above statements,

1— only No. 1 is true.
2— only No. 2 is true.
3— both No. 1 and No. 2 are true.
4— neither No. 1 nor No. 2 is true.

6574. A pressurization controller uses

1—bleed air pressure, outside air temperature, and cabin rate of climb.
2—barometric pressure, cabin altitude, and cabin rate of change.
3—cabin rate of climb, bleed air volume, and cabin pressure.
4—outside air temperature, cabin rate of climb, and cabin pressure.

6575. What unit in a vapor-cycle cooling system serves as a reservoir for the refrigerant?

1—Receiver-dryer.
2—Evaporator.
3—Compressor.
4—Condenser.

6576. What is the condition of the refrigerant as it enters the condenser of a vapor-cycle cooling system?

1—High-pressure liquid.
2—Low-pressure liquid.
3—High-pressure vapor.
4—Low-pressure vapor.

6577. What is the condition of the refrigerant as it enters the evaporator of a vapor-cycle cooling system?

1—High-pressure liquid.
2—Low-pressure vapor.
3—Low-pressure liquid.
4—High-pressure vapor.

6578. What drives the air-conditioning compressor on most small aircraft?

1—An electric motor.
2—Ram air.
3—The aircraft engine through a belt drive.
4—A hydraulic motor.

6579. What is the condition of the refrigerant as it leaves the evaporator of a vapor-cycle cooling system?

1—High-pressure liquid.
2—Low-pressure liquid.
3—Low-pressure vapor.
4—High-pressure vapor.

6580. What is the condition of the refrigerant as it leaves the condenser of a vapor-cycle cooling system?

1—Low-pressure liquid.
2—High-pressure liquid.
3—High-pressure vapor.
4—Low-pressure vapor.

6581. In what position should the bottle be placed when adding liquid freon to a vapor-cycle cooling system?

1—Vertical with the outlet at the top.
2—Vertical with the bottle placed in warm water.
3—Horizontal with the bottle placed in warm water.
4—Vertical with the outlet at the bottom.

6582. What are the normal operating pressures of a freon air-conditioning system with an evaporator temperature of 40 to 50 °F?

1—Low pressure of 100 to 125 PSI and high pressure of 300 to 350 PSI.
2—Low pressure of 10 to 20 PSI and high pressure of 150 to 200 PSI.
3—Low pressure of 65 to 95 PSI and high pressure of 95 to 175 PSI.
4—Low pressure of 20 to 30 PSI and high pressure of 225 to 300 PSI.

6583. When purging a freon air-conditioning system, it is important to release the charge at a slow rate. What is the reason for the slow-rate discharge?

1—Prevent the large amount of freon from contaminating the surrounding atmosphere.
2—Prevent excessive loss of refrigerant oil.
3—Prevent excessive chattering of the expansion valve.
4—Prevent condensation from forming and contaminating the system.

6584. When a vapor-cycle cooling system is not in operation, what is an indication that the system is leaking freon?

1—Oil seepage.
2—Bubbles in the sight glass.
3—Water condensation on a component.
4—An odor of freon in the immediate area.

6585. In an operating vapor-cycle cooling system, if the two lines connected to the expansion valve are essentially the same temperature, what does this indicate?

1—The system is functioning normally.
2—The expansion valve is not metering freon properly.
3—The compressor is pumping too much refrigerant.
4—The system is slightly overcharged.

6586. The purpose of a subcooler in a vapor-cycle cooling system is to

1—augment the cooling capacity during periods of peak demand.
2—cool avionics components that are especially sensitive to heat.
3—provide temporary "quick cooling" for a hot aircraft interior.
4—cool the freon to prevent premature vaporization.

6587. (1) A small amount of water in a vapor-cycle cooling system can freeze in the receiver and stop the entire system operation.

(2) Water in a vapor-cycle cooling system will react with refrigerant to form hydrochloric acid which is highly corrosive to the metal in the system.

Regarding the above statements,

1—only No. 1 is true.
2—only No. 2 is true.
3—both No. 1 and No. 2 are true.
4—neither No. 1 nor No. 2 is true.

6588. What happens to Refrigerant-12 when it passes over an open flame?

1—There is little or no effect.
2—It is consumed by the flame.
3—It changes to phosgene gas.
4—It changes to carbon tetrachloride vapor.

6589. What type of oil is suitable for use in vapor-cycle cooling systems?

1—Low viscosity engine oil.
2—High viscosity engine oil.
3—Special refrigeration oil only.
4—Highly refined synthetic oil, free from impurities with special water absorbing additives.

6590. When it is suspected that the aircraft's oxygen system has developed a leak, the lines and fittings should be checked by

1—coating all suspected areas with a neutral cleaning solvent. Leaks will cause the solvent to spread noticeably.
2—checking the tightness of all fittings with a torque wrench and all lines for cracks by dye check.
3—bubble testing with a testing soap solution manufactured specifically for this purpose.
4—visual inspection for cracks, especially the line flares. If the lines and flares are not cracked and the fittings are properly torqued, no leaks of any consequence will occur.

6591. If oxygen bottle pressure is allowed to drop below a specified minimum, it may cause

1—the pressure reducer to fail.
2—the bottle thermal plug to rupture.
3—the automatic altitude control valve to open.
4—moisture to collect in the bottle resulting in corrosion.

6592. What controls the amount of oxygen delivered to a mask in a continuous-flow oxygen system?

1—Calibrated orifice.
2—Line valve.
3—Pressure reducing valve.
4—Pilot's regulator.

6593. In the diluter-demand oxygen regulator, when does the demand valve operate?

1—When the diluter control is set at normal.
2—When the user demands 100 percent oxygen.
3—When the user breathes.
4—When the cylinder pressure is over 500 PSI.

6594. The primary difference between aviation breathing oxygen and most other types of commercially available compressed oxygen is that aviation breathing oxygen

1—is a mixture of oxygen and other gases essential to normal breathing.
2—has had all water and water vapor removed.
3—contains a measured proportion of hydrogen to reduce the undesirable effects of oxygen on night vision.
4—contains measured amounts of a nontoxic lubricant to prevent fouling of the regulators.

6595. What is used in some oxygen systems to change high cylinder pressure to low system pressure?

1—Pressure reducer valve.
2—Pressure relief valve.
3—Calibrated fixed orifice.
4—Diluter-demand regulator.

6596. If the pressure reducer fails, what prevents high-pressure oxygen from entering the system downstream?

1—Check valve.
2—Cylinder control valve.
3—Pressure relief valve.
4—Manifold control valve.

6597. High-pressure cylinders containing oxygen for aviation use can be identified by their

1—green color and the words "BREATHING OXYGEN" stenciled in 1-inch white letters.
2—yellow color and the words "AVIATOR'S BREATHING OXYGEN" stenciled in 1-inch white letters.
3—yellow color and the words "BREATHING OXYGEN" stenciled in 1-inch white letters.
4—green color and the words "AVIATOR'S BREATHING OXYGEN" stenciled in 1-inch white letters.

6598. (Refer to figure 14.) One hour after an oxygen system was charged for a leakage check, the oxygen pressure gauge read 460 PSI at 63 °F; 6 hours later the temperature was 50 °F. (A 5 PSI change is the maximum allowable in a 6-hour period.) What pressure gauge readings would be acceptable to remain within the allowable limits?

1—455 to 460 PSI.
2—446 to 450 PSI.
3—456 to 460 PSI.
4—445 to 450 PSI.

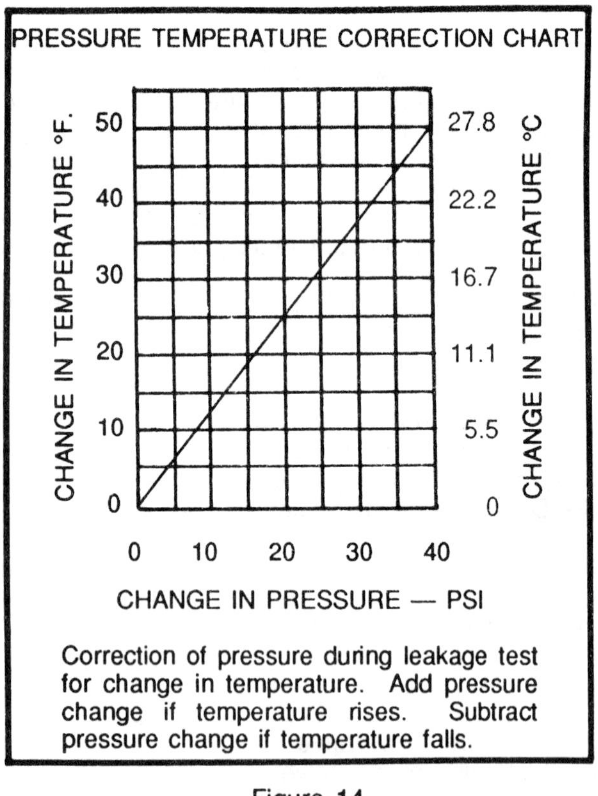

Figure 14

6599. The greatest danger from hypoxia during long exposure at 10,000 feet MSL altitude is

1—headache and fatigue.
2—increased pulse and respiration.
3—blue lips and fingernails.
4—impaired vision and judgment.

6600. In a gaseous oxygen system, which of the following are vented to blow out plugs in the fuselage skin?

1—Pressure relief valves.
2—Filler shutoff valves.
3—Pressure reducer valves.
4—Check valves.

6601. The purpose of pressurizing aircraft cabins is to

(1) create the proper environment for prevention of hypoxia.

(2) permit operation at high altitudes.

Regarding the above statements,

1—only No. 1 is true.
2—only No. 2 is true.
3—both No. 1 and No. 2 are true.
4—neither No. 1 nor No. 2 is true.

6602. (1) Oxygen used in aircraft systems is at least 99.5 percent pure and is practically water free.

(2) Oxygen used in aircraft systems is 99.5 percent pure and is hospital quality.

Regarding the above statements,

1—only No. 2 is true.
2—only No. 1 is true.
3—both No. 1 and No. 2 are true.
4—neither No. 1 nor No. 2 is true.

6603. Oxygen systems in unpressurized aircraft are generally of the

1—continuous-flow and pressure-demand types.
2—continuous-flow type only.
3—pressure-demand type only.
4—portable-bottle type only.

6604. The purpose of the airflow metering aneroid assembly found in oxygen diluter demand regulators is to

1— regulate airflow in relation to oxygen flow when operating in emergency or diluter demand positions.
2— regulate airflow in relation to cabin altitude when in diluter demand position.
3— automatically change to 100 percent oxygen when the aircraft passes the 30,000-foot altitude setting.
4— automatically put the regulator in emergency position if the demand valve diaphragm ruptures.

6605. If a high-pressure oxygen cylinder is to be installed in an airplane, it must meet the specifications of the

1— aircraft manufacturer or the cylinder manufacturer.
2— Interstate Commerce Commission or the Department of Transportation.
3— National Transportation Safety Board or the standards of compressed gas cylinders.
4— Civil Aeronautics Board or the Aviators' Breathing Act.

6606. Before a high-pressure oxygen cylinder is serviced, it must be the correct type and have been

1— hydrostatically tested within the proper time interval.
2— previously installed in an aircraft.
3— approved by the National Transportation Safety Board.
4— inspected by a certificated airframe mechanic.

6607. A contaminated oxygen system is normally purged with

1— castile soap and water.
2— oxygen.
3— compressed air.
4— nitrogen.

6608. How should you determine the amount of oxygen in a portable, high-pressure cylinder?

1— Weigh the cylinder and its contents.
2— Read the pressure gauge mounted on the cylinder.
3— Measure the pressure at the mask.
4— Check the flow indicator during use.

6609. What may be used as a lubricant on oxygen system tapered pipe thread connections?

1— Silicone dielectric compound.
2— Nothing; the threads must be clean and dry.
3— Dry, powdered graphite.
4— Teflon tape.

6610. Which oxygen system uses a molecular sieve as part of the system?

1— Mechanically-separated oxygen system.
2— Liquid oxygen system.
3— Chemical oxygen system.
4— Gaseous oxygen system.

6611. How many of the following are characteristic of a chemical or solid state oxygen system?

A. An adjustable oxygen release rate.
B. A volume storage capacity about three times that of compressed oxygen.
C. The system generators are inert below 400 °F even under severe impact.
D. A distribution and regulating system similar to gaseous oxygen systems.

1— One.
2— Two.
3— Three.
4— Four.

6612. The number of shock mounts required for an original instrument panel installation is determined by the

1— size of the panel.
2— type of the panel.
3— weight of the complete panel unit.
4— number of instruments to be installed.

6613. Which procedure should you use if you find a vacuum-operated instrument glass loose?

1— Mark the case and glass with a white mark.
2— Replace the glass.
3— Reseal the glass in the case.
4— Install another instrument.

6614. Which instruments are connected to the aircraft pitot-static system?

A. Vertical speed indicator.
B. Turn coordinator.
C. Cabin altimeter (pressurized aircraft).
D. Altimeter.
E. Turn-and-slip indicator.
F. Cabin rate-of-change indicator (pressurized aircraft).
G. Airspeed indicator.
H. Directional gyro (air operated).

1—A, D, G.
2—E, G, H.
3—B, C, F.
4—C, F, G.

6615. How many of the following instruments will normally have range markings?

A. Airspeed indicator.
B. Altimeter.
C. Directional gyro.
D. Cylinder head temperature gauge.

1—One.
2—Two.
3—Three.
4—Four.

6616. How would an airspeed indicator be marked to show the best rate-of-climb speed (one engine inoperative)?

1—A white arc.
2—A red radial line.
3—A blue radial line.
4—A green arc.

6617. The green arc on an aircraft temperature gauge indicates

1—the instrument is not calibrated.
2—the desirable temperature range.
3—a low, unsafe temperature range.
4—a high, unsafe temperature range.

6618. What marking on an instrument is used to indicate whether the glass has slipped?

1—Yellow arc.
2—White index mark.
3—Green radial line.
4—Red radial line.

6619. What color paint is used to indicate if the cover glass has slipped?

1—Red.
2—White.
3—Yellow.
4—Green.

6620. Aircraft instruments should be marked and graduated in accordance with

1—the instrument manufacturer's specifications.
2—the specific engine manufacturer's specifications.
3—both the aircraft and engine manufacturers' specifications.
4—the specific aircraft maintenance or flight manual.

6621. Aircraft instrument panels are usually shock-mounted to

1—absorb high-frequency, low-amplitude shocks.
2—absorb low-frequency, high-amplitude shocks.
3—permit the instruments to operate under conditions of major vibration.
4—absorb high-frequency, high-amplitude shocks.

6622. The method of mounting aircraft instruments in their respective panels depends on the

1—instrument manufacturer.
2—design of the instrument case.
3—design of the aircraft fuselage.
4—design of the instrument panel.

6623. How is a flangeless instrument case mounted in an instrument panel?

1—By four machine screws which extend through the instrument panel.
2—By an expanding-type clamp secured to the back of the panel and tightened by a screw from the front of the instrument panel.
3—By a metal shelf separate from and located behind the instrument panel.
4—By press fit into the instrument panel and held in place by friction.

6624. Why are most electrical instruments mounted in iron or steel cases?

1—To avoid damage to the instrument during maintenance.
2—To facilitate removal or installation.
3—To prevent interference from outside magnetic fields.
4—To reduce heat buildup in the instrument.

6625. When installing an instrument in an aircraft, who is responsible for making sure it is properly marked?

1—An authorized inspector.
2—The aircraft owner.
3—The instrument installer.
4—The instrument manufacturer.

6626. Where should you look for the information necessary to determine the required markings of an engine instrument?

1— Engine manufacturer's specifications.
2— Aircraft Specifications or Type Certificate Data Sheets.
3— Specific instrument manufacturer's specifications.
4— Engine and instrument manufacturers' specifications.

6627. A certificated mechanic with airframe and powerplant ratings may

1— perform minor repairs to engine instruments.
2— perform minor repairs and minor alterations to engine instruments.
3— not perform repairs to engine instruments.
4— not perform 100-hour inspections of engine instruments.

6628. A red radial line on the face of an engine instrument indicates

1— normal operating range.
2— caution range.
3— operation is permitted under certain conditions.
4— maximum or minimum safe operating limits.

6629. A certificated mechanic may perform

1— minor alterations to instruments.
2— minor repairs to instruments.
3— 100-hour inspections of instruments.
4— instrument overhaul.

6630. An aircraft instrument panel is electrically bonded to the aircraft structure to

1— allow static electricity buildup.
2— act as a restraint strap.
3— provide current return paths.
4— aid in the panel installation.

6631. How many of the following are controlled by gyroscopes?

A. Attitude indicator.
B. Heading indicator.
C. Turn needle of the turn-and-slip indicator.
D. Angle-of-attack indicator.

1— Four.
2— Three.
3— Two.
4— One.

6632. The lubber line on a directional gyro is used to

1— represent the nose of the aircraft.
2— align the instrument glass in the case.
3— represent the wings of the aircraft.
4— indicate true north as opposed to magnetic north.

6633. An aircraft magnetic compass is swung at specified operating intervals in order to determine the

1— accuracy of the lubber line.
2— compass precession.
3— compass variation.
4— compass deviation.

6634. The operating mechanism of most hydraulic pressure gauges is

1— a Bourdon tube.
2— an airtight bellows.
3— an airtight diaphragm.
4— an evacuated bellows filled with an inert gas to which suitable arms, levers, and gears are attached.

6635. What is the fixed reference marker attached to the compass bowl of a magnetic compass called?

1— Reeder line.
2— Lubber line.
3— Card line.
4— Pole line.

6636. (1) Aircraft instruments are color-coded to direct attention to approaching operating difficulties.

(2) Aircraft instruments range markings are not specified by Federal Aviation Regulations but are standardized by aircraft manufacturer agreement.

Regarding the above statements,

1— only No. 1 is true.
2— only No. 2 is true.
3— neither No. 1 nor No. 2 is true.
4— both No. 1 and No. 2 are true.

6637. When swinging a magnetic compass, the compensators are adjusted to correct for

1— magnetic influence deviation.
2— small magnetic disturbances.
3— magnetic variations.
4— card oscillations.

6638. What will be the result if the instrument static pressure line becomes disconnected inside a pressurized cabin during cruising flight?

1— The altimeter and airspeed indicator will both read low.
2— The altimeter and airspeed indicator will both read high.
3— The altimeter will read high and the airspeed indicator will read low.
4— The altimeter will read low and the airspeed indicator will read high.

6639. Which statement relating to instruments and instrument systems is true?

1— The rate-of-climb indicator uses pitot pressure.
2— Apply suction when leak testing pitot lines.
3— Apply pressure when leak testing static lines.
4— The airspeed indicator uses both pitot and static pressure.

6640. The maximum deviation (during level flight) permitted in a magnetic direction indicator installed in an aircraft is

1— 4°.
2— 6°.
3— 8°.
4— 10°.

6641. Magnetic compass bowls are filled with a liquid to

1— retard precession of the float.
2— reduce deviation errors.
3— counteract temperature and attitude changes.
4— dampen the oscillation of the float.

6642. Instrument static system leakage can be detected by observing the rate of change in indication of the

1— airspeed indicator after suction has been applied to the static system to cause a prescribed equivalent airspeed to be indicated.
2— altimeter after pressure has been applied to the static system to cause a prescribed equivalent altitude to be indicated.
3— airspeed indicator after pressure has been applied to the static system to cause a prescribed equivalent airspeed to be indicated.
4— altimeter after suction has been applied to the static system to cause a prescribed equivalent altitude to be indicated.

6643. The maximum altitude loss permitted during an unpressurized aircraft instrument static pressure system integrity check is

1— 150 feet in 2 minutes.
2— 50 feet in 1 minute.
3— 75 feet in 2 minutes.
4— 100 feet in 1 minute.

6644. Which statement regarding an aircraft instrument vacuum system is true?

1— Dry-type vacuum pumps with carbon vanes are very susceptible to damage from solid airborne particles and must take in only filtered air.
2— Vacuum systems are generally more effective at high altitudes than positive pressure systems.
3— A restrictor valve is generally installed in the air-inlet line of the attitude indicator (artificial horizon).
4— If the air inlet to each vacuum instrument is connected to a common atmospheric pressure manifold, the system generally will be equipped with individual instrument filters only.

6645. When an aircraft altimeter is set at 29.92" Hg on the ground, the altimeter will read

1— pressure altitude.
2— density altitude.
3— field elevation.
4— true altitude.

6646. Which of the following instrument discrepancies could be corrected by an aviation mechanic?

A. Red line missing.
B. Case leaking.
C. Glass cracked.
D. Mounting screws loose.
E. Case paint chipped.
F. Leaking at line B nut.
G. Will not zero out.
H. Fogged.

1— A, F, D.
2— C, D, E, F.
3— A, C, E.
4— A, D, E, F.

6647. Which of the following instrument discrepancies would require replacement of the instrument?

A. Red line missing.
B. Case leaking.
C. Glass cracked.
D. Mounting screws loose.
E. Case paint chipped.
F. Leaking at line B nut.
G. Will not zero out.
H. Fogged.

1—A, D, E, F.
2—B, C, G, H.
3—A, D, F, G.
4—A, C, E, H.

6648. Which of the following instrument conditions is acceptable and would not require correction?

A. Red line missing.
B. Case leaking.
C. Glass cracked.
D. Mounting screws loose.
E. Case paint chipped.
F. Leaking at line B nut.
G. Will not zero out.
H. Fogged.

1—A.
2—D.
3—E.
4—None.

6649. A barometric altimeter indicates pressure altitude when the barometric scale is set at

1—29.92 inches of mercury.
2—14.7 millibars.
3—14.7 inches of mercury.
4—field elevation.

6650. A Bourdon tube instrument may be used to indicate

A. Pressure.
B. Temperature.
C. Position.
D. Quantity.

1—A and B.
2—C and D.
3—A only.
4—B and D.

6651. A turn-and-bank instrument indicates

1—the longitudinal attitude of the aircraft during climb.
2—trim and serves as an emergency source of bank information in case the attitude gyro fails.
3—the longitudinal attitude of the aircraft during descent.
4—the need for corrections in pitch and bank anytime the aircraft deviates from a preselected attitude.

6652. A typical deicing pressure gauge case

1—is considered to be a sealed unit.
2—houses a Bourdon tube and diaphragm.
3—contains a magnet and sector gear.
4—is vented to keep the interior at atmospheric pressure.

6653. Thermocouple leads

1—are designed for a specific installation and may not be altered.
2—may be adjusted in length to fit any installation.
3—may be installed with either lead to either post of the indicator.
4—may be repaired using solderless connectors.

6654. A synchro transmitter is connected to a synchro receiver

1—mechanically through linkage.
2—magnetically through highly permeable cores.
3—electrically with wires.
4—for appearance only since there is no physical connection between the two.

6655. The operation of the angle-of-attack indicating system is based on detection of differential pressure at a point where the airstream flows in a direction

1—not parallel to the true angle of attack of the aircraft.
2—parallel to the angle of attack of the aircraft.
3—parallel to the angle of incidence.
4—parallel to the longitudinal axis of the aircraft.

6656. Turbine engine exhaust gas temperatures are measured by using

1—iron/constantan thermocouples.
2—electrical resistance thermometers.
3—chromel/alumel thermocouples.
4—ratiometer electrical resistance thermometers.

6657. Fuel flow transmitters are designed to transmit data

1— mechanically.
2— electrically.
3— visually.
4— utilizing fluid power.

6658. Which of the following causes of aircraft magnetic compass inaccuracies may be compensated for by mechanics?

1— Deviation.
2— Magnetic compass dip error.
3— Magnetic compass current.
4— Variation.

6659. Who is authorized to repair an aircraft instrument?

1— A certificated mechanic with airframe and powerplant ratings.
2— A certificated repair station approved for that class instrument.
3— An appropriately rated airframe repair station.
4— A certificated mechanic holding an Inspection Authorization.

6660. What does a reciprocating engine manifold pressure gauge indicate when the engine is not operating?

1— Zero pressure.
2— The differential between the manifold pressure and the atmospheric pressure.
3— Corrected differential pressure.
4— The existing atmospheric pressure.

6661. (1) The angle-of-attack indicating system measures the angle between the local airflow around the direction detector and the fuselage reference plane.

(2) The operation of the angle-of-attack system is based on the measurement of pressure differential at a point where the airstream is not flowing parallel to the true angle of attack.

Regarding the above statements,

1— both No. 1 and No. 2 are true.
2— only No. 1 is true.
3— only No. 2 is true.
4— neither No. 1 nor No. 2 is true.

6662. The minimum requirements for testing and inspection of instrument static pressure systems required by FAR Section 91.171 are contained in

1— Type Certificate Data Sheets.
2— Technical Standard Orders.
3— AC 43.13-1A.
4— FAR Part 43, appendix E.

6663. Which condition would be most likely to cause excessive vacuum in a vacuum system?

1— Vacuum pump overspeed.
2— Vacuum relief valve improperly adjusted.
3— Venturi-tube icing.
4— Vacuum relief valve spring weak.

6664. What is the primary purpose of an autopilot?

1— To relieve the pilot of control of the aircraft during long periods of flight.
2— To provide a secondary system of aircraft guidance.
3— To fly a more precise course for the pilot.
4— To obtain the navigational aid necessary for extended overwater flights.

6665. Which of the following provides manual maneuverability of the aircraft while the autopilot is engaged?

1— Servo-amplifier.
2— Attitude indicator.
3— Directional gyro indicator.
4— Flight controller.

6666. In an autopilot, which signal nullifies the input signal to the ailerons?

1— Displacement signal.
2— Course signal.
3— Rate signal.
4— Followup signal.

6667. In which control element of an autopilot system is an attitude indicator?

1— Command.
2— Sensing.
3— Computer.
4— Input.

6668. What is the operating principle of the sensing device used in an autopilot system?

1— The reaction of the force 90° away from the applied force in the direction of gyro rotation.
2— The relative motion between a gyro and its supporting system.
3— The rate of change of motion between the gyro gimbal rings and the aircraft.
4— The interaction of the applied force and the rigidity of the gyro.

6669. What is the purpose of displacement followup in an autopilot system?

1— Develop a signal proportional to the amount of flight deviation.
2— Transmit a rate-of-change signal proportional to the rate of recovery.
3— Develop and transmit a negative feedback signal.
4— Transmit a signal to the controller when the aircraft is on course.

6670. What component of an autopilot system applies torque to the control surfaces of an aircraft?

1— Servo.
2— Controller.
3— Gyro.
4— Computer.

6671. What is the main purpose of a servomotor in an autopilot system?

1— Correct for displacement of the aircraft about its axis.
2— Change mechanical energy to electrical energy.
3— Move the control surface as commanded.
4— Drive the control surface back to the streamlined position.

6672. Which channel of an autopilot detects changes in pitch attitude of an aircraft?

1— Elevator channel.
2— Aileron channel.
3— Rudder channel.
4— Roll channel.

6673. The elevator channel of an autopilot controls the aircraft about which axis of rotation?

1— Roll axis.
2— Longitudinal axis.
3— Lateral axis.
4— Yaw axis.

6674. What component is the sensing device in an electromechanical autopilot system?

1— Servo.
2— Turn and bank.
3— Gyro.
4— Controller.

6675. A fully integrated autopilot controls the aircraft around how many axes?

1— One.
2— Two.
3— Three.
4— Four.

6676. Dutch roll, a combination yawing and rolling oscillation that affects many sweptwing aircraft, is counteracted with

1— a flight director system.
2— a pitch damper system.
3— an aileron damper system.
4— a yaw damper system.

6677. When operationally checking an autopilot system on the ground, after the aircraft's main power has been switched on, the autopilot should be engaged

1— as soon as possible.
2— only after the gyros come up to speed and the amplifier warms up.
3— whenever the operator desires.
4— for only a few minutes at a time.

6678. What protects radio equipment from vibration damage?

1— Shock mounts.
2— Aluminum-alloy jumpers.
3— Self-locking hold-down clamps and snap slides.
4— Doubler plates around the mounting points.

6679. (1) Use solder to attach bonding jumpers on radio equipment.

(2) Radio equipment is bonded to the aircraft in order to provide a low-impedance ground and to minimize radio interference from static electrical charges.

Regarding the above statements,

1— only No. 1 is true.
2— both No.1 and No. 2 are true.
3— neither No. 1 nor No. 2 is true.
4— only No. 2 is true.

6680. Which of the following must be displayed in an aircraft equipped with a two-way radio?

1— Radio station license.
2— Certificate of TSO authorization.
3— Restricted radiotelephone operator's permit.
4— Certificate of minimum performance standards.

6681. Which of the following issues the license required to operate an aircraft radio transmitter?

1— Department of Defense.
2— Department of Interior.
3— Federal Aviation Administration.
4— Federal Communications Commission.

6682. What is the minimum license a person must hold to be permitted to operate an aircraft transmitter?

1— First-class radiotelephone.
2— Second-class radiotelephone.
3— Airframe and powerplant mechanic.
4— Restricted radiotelephone.

6683. When installing coaxial cable, it should be secured firmly along its entire length

1— at 1-foot intervals.
2— wherever the cable sags.
3— at 2-foot intervals.
4— at 3-foot intervals.

6684. The emergency locator transmitter (ELT) battery

1— replacement date must be marked on the outside of the transmitter.
2— must be a dry-cell type.
3— must be replaced annually.
4— replacement date must be computed from the date of installation.

6685. An emergency locator transmitter (ELT) battery must be capable of furnishing power for signal transmission for at least

1— 12 hours.
2— 24 hours.
3— 36 hours.
4— 48 hours.

6686. What is the preferred location of an emergency locator transmitter (ELT)?

1— At the approximate intersection of the lateral and longitudinal axes of the aircraft.
2— Readily accessible to the pilot or a member of the flightcrew when the aircraft is in flight.
3— As far forward as possible, but aft of the cabin section.
4— As far aft as possible, but forward of the vertical fin.

6687. An emergency locator transmitter (ELT) is normally activated by an inertial switch or equivalent mechanism if subjected to a force of a prescribed intensity and duration. It must activate when the force is applied

1— parallel to the longitudinal axis of the aircraft.
2— parallel to the vertical axis of the aircraft.
3— in any direction relative to the aircraft axes.
4— parallel to the lateral axis of the aircraft.

6688. How may the battery replacement date for an emergency locator transmitter (ELT) be determined?

1— By removing the batteries and testing them under a measured load to determine if 50 percent of the useful life remains.
2— By inspecting the aircraft maintenance records to determine installation date. ELT batteries must be replaced each 12 months.
3— By observing the battery replacement date marked on the outside of the transmitter.
4— By activating the transmitter and measuring the signal strength.

6689. What is the most common power source for an emergency locator transmitter (ELT)?

1— A self-contained battery.
2— A direct unfused connection to the aircraft electrical power system.
3— A special interconnection to the aircraft battery which allows current to flow in only one direction.
4— An isolating transformer connected to the normal aircraft electrical power system.

6690. How may the operation of an installed emergency locator transmitter (ELT) be verified during aircraft inspection?

1—By moving the deactivating switch from the DISARM position to the ARM position while monitoring the civil emergency frequency with a communications receiver.
2—By grounding the ELT antenna through a specially constructed meter and activating the transmitter. The meter reading should be within the limits established by the manufacturer.
3—By visually inspecting the transmitter installation and checking its power supply.
4—By tuning a communications receiver to the civil emergency frequency and activating the ELT momentarily.

6691. Which of the following conditions requires replacement of emergency locator transmitter (ELT) batteries?

1—When the transmitter has a total of 30 minutes of operation.
2—Anytime the transmitter has operated continuously for 30 minutes or has a total of 45 minutes of operation.
3—When the batteries reach 50 percent of their useful (shelf) life as established by the manufacturer.
4—When the ambient temperature has risen to 110 °F for 6 or more hours.

6692. Static dischargers are installed on aircraft to bleed off static electricity. Elimination of static electricity is desired in order to

1—reduce radio receiver interference.
2—prevent opposition and surges in the aircraft's generated power system.
3—keep the passengers and/or crew from experiencing static electricity shocks.
4—prevent false indications on the aircraft's instruments.

6693. When an antenna is installed, it should be fastened

1—to the primary structure at the aircraft's directional pivotal point.
2—to the primary structure at the approximate intersection of the three aircraft axes.
3—with a reinforcing doubler on each side of the aircraft skin.
4—so that loads imposed are transmitted to the aircraft structure.

6694. After an automatic direction finding antenna has been installed,

1—the antenna must be grounded.
2—the loop must be calibrated.
3—the extra length of wire between the loop and receiver must be removed.
4—the transceiver must be compensated.

6695. Doublers are used when antennas are installed

1—to eliminate antenna vibration.
2—to reduce aircraft flutter.
3—to prevent oil canning of the skin.
4—to reinstate the structural strength of the aircraft skin.

6696. One antenna can be used for the radio range and standard broadcast bands in light aircraft because the

1—two ranges are close together.
2—antenna is omnidirectional.
3—antenna length may be electronically adjusted.
4—quadrantal error is minimized.

6697. What characteristics of the installation of a rigid antenna on a vertical stabilizer should be evaluated?

1—Polarization and impedance.
2—Impedance and interference.
3—Flutter and vibration.
4—Sensitivity and response.

6698. A gasket or sealant is used between the antenna mast and fuselage skin

1—to prevent the entry of moisture.
2—so the attachment studs may be drawn tighter.
3—for aircraft pressurization only.
4—to prevent abrasion between the antenna mast and fuselage skin.

6699. The preferred location of a VOR antenna on light aircraft is

1—on the bottom of the fuselage and as far forward as possible.
2—any convenient location on the top of the fuselage.
3—on top of the cabin with the apex on the V pointing forward.
4—on top of the vertical stabilizer.

6700. The purpose of a localizer is to

1— locate lost airplanes.
2— set the airplane on the proper approach angle to the runway.
3— indicate the distance the airplane is from the end of the runway.
4— align the airplane with the center of the runway.

6701. (Refer to figure 15.) What is the approximate drag load on an antenna with a frontal area of .125 square feet installed on an aircraft with a speed of 225 MPH?

1— 2.069 pounds.
2— 2.073 pounds.
3— 2.080 pounds.
4— 2.059 pounds.

6702. (Refer to figure 15.) What is the approximate drag load on an antenna with a frontal area of .137 square feet installed on an aircraft with a speed of 275 MPH?

1— 3.387 pounds.
2— 3.932 pounds.
3— 3.741 pounds.
4— 3.592 pounds.

$$D = .000327AV^2$$

Figure 15

6703. An antenna is a special type of electrical circuit designed to radiate and receive

1— electromagnetic energy.
2— audible signals.
3— visual signals.
4— subharmonic frequencies.

6704. A DME antenna should be located in a position on the aircraft that will

1— not be blanked by the wing when the aircraft is banked.
2— allow an antenna functional check to be made on the ground without a DME test set.
3— permit interruptions in DME operation.
4— eliminate the possibility of the DME locking on a station.

6705. When bending coaxial cable, the bend radius should be at least

1— 5 times the diameter of the cable.
2— 10 times the diameter of the cable.
3— 15 times the diameter of the cable.
4— 20 times the diameter of the cable.

6706. When installing a DME antenna, it should be aligned with the

1— angle of decalage.
2— null position.
3— angle of incidence.
4— centerline on the airplane.

6707. (Refer to figure 16.) Which of the antennas shown is a typical DME antenna?

1— A.
2— B.
3— C.
4— D.

6708. (Refer to figure 16.) Which of the antennas shown is a typical glideslope antenna?

1— A.
2— B.
3— C.
4— D.

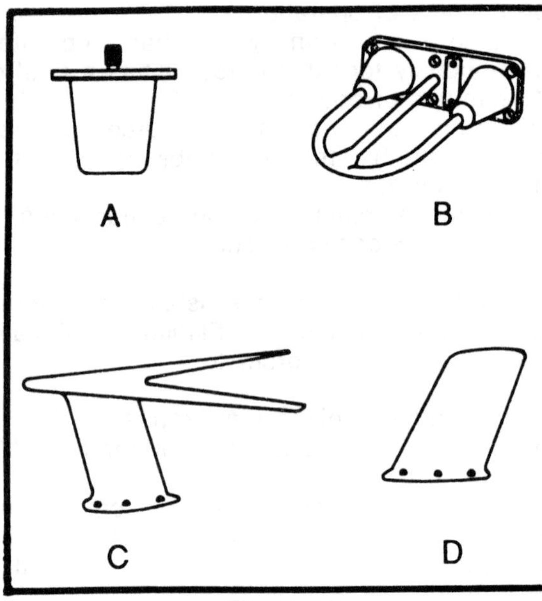

Figure 16

6709. The addition of avionics and associated antenna systems forward of the CG limit will affect

1— empty weight and useful load.
2— CG limits and useful load.
3— useful load and maximum gross weight.
4— maximum gross weight and datum.

6710. To communicate with ground control from an aircraft, you would use

1— a VOR receiver.
2— an ADF.
3— a VHF transceiver.
4— an HF transmitter.

6711. How much clearance from the seat bottom is required when installing radio equipment under a seat?

1— 3 inches with the seat unoccupied.
2— 3 inches with the seat occupied.
3— No set minimum as long as the equipment receives adequate cooling and damage protection.
4— 1 inch with the seat occupied and subjected to maximum downward seat spring deflection.

6712. The purpose of a glideslope system is to

1— provide for automatic altitude reporting to air traffic control.
2— indicate the distance the airplane is from the end of the runway.
3— assist the pilot in making the correct angle of incidence approach to the runway.
4— assist the pilot in making a correct angle of descent to the runway.

6713. If the maximum takeoff weight for a transport category aircraft is 105 percent of the maximum landing weight, what system is required?

1— Fuel jettison.
2— Fuel injection.
3— Crossfeed bypass.
4— Fuel transfer.

6714. Fuel jettisoning is usually accomplished

1— through a common manifold and outlet in each wing.
2— by gravity flow into the outboard wing tanks and overboard through a common outlet in each wing.
3— by pump pressure into the crossfeed manifold and overboard through the vent lines.
4— through individual outlets for each tank.

6715. Which of the following is a requirement for fuel jettison systems?

1— The system and its operation must be free of fire hazards.
2— The time required to jettison fuel must not exceed 10 minutes.
3— The fuel must be discharged from outlets at the tail of the aircraft.
4— The fuel must be discharged from outlets at the wingtips.

6716. (1) The fuel jettison valve must be designed to allow flight personnel to close the valve during any part of the jettisoning operation.

(2) During the fuel jettisoning operation, the fuel must discharge clear of any part of the airplane.

Regarding the above statements,

1— both No. 1 and No. 2 are true.
2— only No. 2 is true.
3— neither No. 1 nor No. 2 is true.
4— only No. 1 is true.

6717. A fuel jettison system consists of

1— filters, switches, valves, dump chutes, and chute operating mechanisms.
2— lines, valves, dump chutes, and chute operating mechanisms.
3— tanks, filters, valves, lines, dump chutes, and chute operating mechanisms.
4— flowmeters, filters, valves, lines, dump chutes, and chute operating mechanisms.

6718. Which of the following is employed to maintain lateral stability when jettisoning fuel?

1— Two separate independent systems.
2— Crossfeed system.
3— Two interconnected systems.
4— Equalizer system.

6719. A fuel jettison system is required if the maximum takeoff weight exceeds the maximum landing weight on

1— transport category aircraft only.
2— aircraft that weigh over 12,500 pounds only.
3— both transport category and general aviation aircraft.
4— general aviation aircraft only.

6720. Fuel is moved overboard in most fuel jettison systems by

1— engine-driven fuel pumps.
2— boost pumps.
3— gravity.
4— gravity and engine-driven fuel pumps.

6721. Fuel jettisoning past the limits prescribed by Federal Aviation Regulations is prevented

1— by closely monitoring the fuel quantity and turning off the fuel dump switch(es).
2— by dump limit valves or a low level circuit.
3— by a variable low level circuit or a variable dump limit valve system that is preset prior to each flight.
4— by standpipes in the fuel tanks.

6722. Which procedure must be followed when defueling aircraft with sweptback wings?

1— Defuel all the tanks at one time.
2— Defuel the fuselage tanks last.
3— Defuel the inboard wing tanks first.
4— Defuel the outboard wing tanks first.

6723. (Refer to figure 17.) What is the purpose of the pump crossfeed valve?

1— Balance the fuel in the tanks.
2— Allow operation of engines from one tank.
3— Allow operation of the left engine when the right fuel-boost pump is inoperative.
4— Allow operation of both engines.

6724. Why are crossfeed systems used in multiengine aircraft fuel systems?

1— To permit dumping of excess fuel.
2— To reduce number of fuel lines.
3— To maintain a balanced fuel condition.
4— To reduce refueling time.

6725. What is the primary purpose of the crossfeed system?

1— To allow the feeding of any engine from any tank.
2— To allow removal of residual fuel.
3— To allow the feeding of fuel from one tank for defueling.
4— To provide automatic refueling of a tank to any desired level.

6726. Fuel system components must be bonded and grounded in order to

1— drain off static charges.
2— prevent stray currents.
3— identify the components.
4— stabilize the units.

6727. A typical large aircraft fuel manifold system allows how many of the following?

A. All tanks can be serviced through a single connection.
B. Any engine can be fed from any tank.
C. All engines can be fed from all tanks simultaneously.
D. A damaged tank can be isolated from the rest of the fuel system.

1— One.
2— Two.
3— Three.
4— Four.

6728. The use of turbine fuels in aircraft has resulted in some problems not normally associated with aviation gasolines. One of these problems is

1— increasing viscosity of fuel as fuel temperature is lowered.
2— a health hazard due to easy and rapid absorption through the skin.
3— high vapor pressure.
4— microbial contaminants.

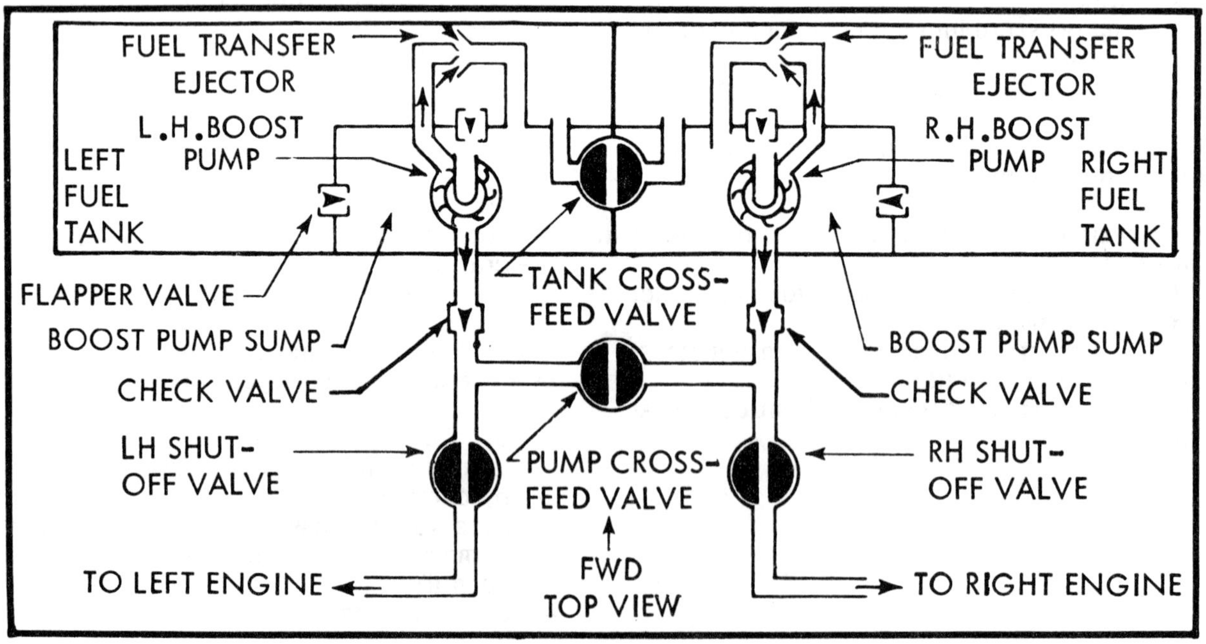

Figure 17

6729. What is used in many aircraft to prevent bubbles in the fuel after it leaves the tank when atmospheric pressure is lower than fuel vapor pressure?

1— Air-fuel separators.
2— Fuel additives.
3— Boost pumps.
4— Restrictor valves.

6730. What is one precaution that must be observed during fueling operations?

1— All outside electrical sources must be disconnected from the aircraft.
2— Fuel to be used must be appropriately identified.
3— All electrical switches must be in OFF position.
4— Fuel must be filtered through a chamois.

6731. Before fueling an aircraft by using the pressure fueling method, what important precaution should be observed?

1— The truck pump pressure must be correct for that refueling system.
2— The truck pump pressure must be adjusted for minimum filter pressure.
3— The hose must be connected before grounding.
4— The aircraft's electrical system must be on to indicate quantity gauge readings.

6732. What flight safety related advantage does a pressure fueling system provide?

1— Eliminates aircraft skin damage.
2— Increases the chances of static electricity igniting fuel vapors.
3— Reduces the chances for fuel contamination.
4— Reduces the time required for fueling.

6733. Aircraft pressure fueling systems instructional procedures are normally placarded on the

1— flightcrew checklist.
2— fuel control panel access door.
3— lower wing surface adjacent to the access door.
4— ground crew checklist.

6734. Why is the pressure fueling method more practical for large aircraft than the over-the-wing method?

1— It reduces training time for fuel servicing personnel.
2— It saves time by providing a single point to fuel the entire aircraft.
3— It saves fuel by reducing the chances for overfills.
4— It reduces the possibility of structural damage to bladder-type fuel cells.

6735. Which of the following may be used to repair most integral fuel tanks?

1— Welding.
2— Brazing.
3— Soldering.
4— Riveting.

6736. How is the outlet fuel pressure regulated on a submerged, single-speed, centrifugal-type fuel pump?

1— By the engine-driven pump's design and internal clearance.
2— By the depth the pump is submerged in fuel.
3— By the first check valve downstream from the pump.
4— By the pump's design and internal clearances.

6737. What is one purpose of a fuel tank vent?

1— To maintain atmospheric pressure.
2— To decrease fuel vapor pressure.
3— To prevent condensation at temperature change.
4— To decrease tank internal air pressure.

6738. What publication defines the test pressures to be followed when testing a repaired fuel tank of a small aircraft?

1— Federal Aviation Regulations.
2— Manufacturer's service letters.
3— AC 43.13-1A.
4— AC 43.13-2A.

6739. If it is necessary to enter an aircraft's fuel tank, which procedure should be avoided?

1— Continue purging the tank during the entire work period.
2— Station an assistant outside the fuel tank access to perform rescue operations if required.
3— Conduct the defueling and tank purging operation in an air-conditioned building.
4— Before entering the tank, touch a static discharge plate.

6740. What is the recommended practice for cleaning a fuel tank before welding?

1— Immerse in a 5 percent solution of nitric acid.
2— Purge the tank with air.
3— Wash inside of tank with clean water.
4— Steam clean the tank interior.

6741. An aircraft's integral fuel tank is

1— removable from the aircraft.
2— usually located in the bottom of the fuselage.
3— a part of the aircraft structure.
4— a self-sealing tank.

6742. Which gas is used for purging an aircraft fuel tank?

1— Hydrogen.
2— Carbon dioxide.
3— Oxygen.
4— Carbon monoxide.

6743. Why is the main fuel strainer located at the lowest point in the fuel system?

1— It eliminates the need for fuel tank sumps.
2— It traps any small amount of water that may be present in the fuel system.
3— It provides a drain for residual fuel.
4— It filters and traps all micro-organisms that may be present in the fuel system.

6744. The purpose of a diaphragm in a vane-type fuel pump is to

1— equalize fuel pressure at all speeds.
2— vary fuel pressure according to throttle setting.
3— prevent fuel pressure from exceeding venturi pressure.
4— compensate fuel pressures to altitude changes.

6745. When moving the mixture control on a normally operating engine into the idle cutoff position, engine RPM should

1— immediately began to drop.
2— slightly increase before the engine starts to die.
3— slightly decrease and then drop rapidly.
4— remain the same until the cutoff is effected, then drop rapidly.

6746. Entrained water in aviation turbine fuel is a hazard because of its susceptibility to freezing as it passes through the filters. What is a common method of preventing this hazard?

1— Deicing fluid in the fuel.
2— Micromesh fuel strainers.
3— High-velocity fuel pumps.
4— Anti-icing fuel additives.

6747. What is one of the first items to be checked when troubleshooting a rough running fuel-injected engine?

1— Excessive fuel pressure.
2— Inadequate fuel flow.
3— Clogged air screen.
4— Clogged fuel nozzles.

6748. The presence of fuel stains around a fuel nozzle would indicate

1— too much fuel pressure.
2— excessive airflow across the venturi.
3— clogged fuel nozzle.
4— loose manifold.

6749. What should be used to inert an integral fuel tank before attempting repairs?

1— CO_2.
2— Water.
3— Steam.
4— Compressed air.

6750. What should be used to remove flux from an aluminum tank after welded repairs?

1— Soft brush and warm water.
2— 5 percent solution of hydrochloric acid.
3— 5 percent solution of nitric or sulfuric acid.
4— Mild solution of soap and warm water.

6751. What method would be used to check for internal leakage of a fuel valve without removing the valve from the aircraft?

1— Place the valve in the OFF position, drain the strainer bowl, and with boost pump on, watch to see if fuel flows to the strainer bowl.
2— Remove fuel cap(s), turn boost pump(s) on, and watch for bubbling in the tanks.
3— Disconnect valve inlet line and observe for fuel leakage.
4— Apply regulated air pressure on the downstream side of the fuel pump and listen for air passing through the valve.

6752. Why are jet fuels more susceptible to water contamination than aviation gasoline?

1—Jet fuel has a higher viscosity than gasoline.
2—Condensation is greater because of the rapid burn-off of jet fuels.
3—Jet fuel is lighter than gasoline; therefore, water is more easily suspended.
4—Processing and handling of jet fuel is less stringent than gasoline.

6753. When installing a rigid fuel line, 1/2 inch in diameter, at what intervals should the line be supported?

1—24 inches.
2—36 inches.
3—12 inches.
4—16 inches.

6754. The probe of a capacitance-type fuel level gauge is essentially a

1—float-actuated variable condenser.
2—condenser with fuel and air acting as one plate.
3—condenser with fuel and air acting as a dielectric.
4—float-actuated variable resistor.

6755. The capacitance-type (electronic) fuel quantity indicator

1—has no moving parts in the tank.
2—has two tubes separated by a mica dielectric in the tank.
3—is not accurate when fuel temperature exceeds 100 °F.
4—measures the amount of fuel in each tank and reads in gallons.

6756. What type of remote-reading fuel quantity indicating system has several probes installed in each fuel tank?

1—Mechanical.
2—Electromechanical.
3—Electronic.
4—Direct reading.

6757. Which aircraft fuel quantity indicating system incorporates a signal amplifier?

1—Electronic.
2—Sight glass.
3—Mechanical.
4—Electrical.

6758. A drip gauge may be used to measure

1—the amount of fuel in the tank.
2—fuel selector valve leakage.
3—system leakage with the system shut down.
4—fuel pump diaphragm leakage.

6759. The electronic-type fuel quantity indicating system consists of a bridge circuit,

1—an amplifier, an indicator, and a tank unit.
2—a tank, an amplifier, and an indicator.
3—a tank unit, a tank, and an amplifier.
4—an indicator, a tank unit, and a tank.

6760. A probe or a series of probes is used in what kind of fuel quantity indicating system?

1—Magnesyn.
2—Selsyn.
3—Capacitor.
4—Synchro.

6761. The electronic-type fluid quantity indicating system is more accurate in measuring fuel level because

1—it measures in gallons and converts to pounds.
2—only one probe and one indicator are necessary for multiple tank configurations.
3—aircraft attitude has no effect on fluid quantity indication.
4—it measures by weight instead of gallons.

6762. One advantage of electrical and electronic fuel quantity indicating systems is that

1—the indicators are calibrated in gallons; therefore, no conversion is necessary.
2—only one transmitter and one indicator are needed regardless of the number of tanks.
3—several fuel tank levels can be read on one indicator.
4—once calibrated, no further adjustment or calibration is required.

6763. A fuel totalizer is a component which

1—measures the amount of fuel being delivered to each engine.
2—shows the amount of fuel in any given tank.
3—indicates the total rate at which all engines are consuming fuel.
4—indicates the amount of fuel in all the fuel tanks.

6764. What is the dielectric (nonconducting material) of the condenser in a capacitance-type fuel quantity indicating system?

1— Outer shell of the condenser.
2— Wire coil external to the tank.
3— Fuel in the tank.
4— Fuel and air (vapor) above the fuel.

6765. A capacitance-type fuel quantity indicating system measures fuel in

1— pounds.
2— pounds per hour.
3— gallons per hour.
4— gallons.

6766. What are the four general types of fuel quantity gauges?

A. Sight glass.
B. Mechanical.
C. Electrical.
D. Electronic.
E. Bourdon tube.
F. Vane-type transmitter.
G. Litmus indicator.
H. Direct-reading static pressure type.

1— A,B,C,D.
2— A,C,F,H.
3— B,C,E,G.
4— E,F,G,H.

6767. How does temperature affect fuel weight?

1— Cold fuel is heavier per gallon.
2— Warm fuel is heavier per gallon.
3— Temperature has no effect.
4— Temperature affects fuel weight at sea level but has no effect at cruising altitude.

6768. One advantage of electrical and electronic fuel quantity indicating systems is that the indicator

1— can be located any distance from the tank.
2— indicates the fuel quantity for one tank only.
3— has no movable devices.
4— is always calibrated in gallons; therefore, no conversion is necessary.

6769. When fuel quantity is measured in pounds instead of gallons, the measurement will be more accurate because fuel volume

1— varies with temperature change.
2— remains unchanged when temperature decreases.
3— increases when temperature decreases.
4— remains unchanged when temperature increases.

6770. An electrical-type fuel quantity system consists of an indicator in the cockpit and a

1— float-operated transmitter installed in the tank.
2— glass or plastic tube placed on the same level as the tank.
3— float resting on the surface of the tank.
4— float-operated receiver installed in the tank.

6771. What is the purpose of a float-operated transmitter installed in a fuel tank?

1— It sends an electric signal to the fuel quantity indicator.
2— It senses the total amount of fuel density.
3— It senses the dielectric qualities of fuel and air in the tank.
4— It senses fuel pressure.

6772. In an electronic-type fuel quantity indicating system, the tank transmitter is

1— an electric condenser.
2— a variable resistor.
3— an electric amplifier.
4— a variable capacitor.

6773. What must each fuel quantity indicator be calibrated to read during level flight when the quantity of fuel remaining is equal to the unusable fuel supply?

1— The total unusable fuel quantity.
2— The unusable fuel quantity in each tank.
3— Both the total unusable fuel quantity and the unusable fuel quantity in each tank.
4— Zero.

6774. What unit would be adjusted to change the fuel pressure warning limits?

1— Fuel flowmeter bypass valve.
2— Engine fuel pump bypass valve.
3— Pressure-sensitive mechanism.
4— Fuel pressure relief valve.

6775. Select one means of controlling the fuel temperature on turbine-powered aircraft.

1— Electrically at the fuel filter.
2— Engine lubricating oil at the fuel filter.
3— Engine bleed air around the fuel tank.
4— Engine bleed air to a heat exchanger.

6776. What is the purpose of flapper-type check valves in integral fuel tanks?

1— To prevent overpressurization during refueling operations.
2— To allow defueling of the tanks by suction.
3— To prevent fuel from flowing away from the boost pumps.
4— To allow the engine-driven pumps to draw fuel directly from the tank if the boost pump fails.

6777. What unit is generally used to actuate the fuel pressure warning system?

1— Fuel flowmeter.
2— Pressure-sensitive mechanism.
3— Engine fuel pump bypass valve.
4— Fuel pressure gauge.

6778. What method is used on turbine-powered aircraft to determine when the condition of the fuel is approaching the danger of forming ice crystals?

1— Fuel pressure warning.
2— Fuel pressure gauge.
3— Fuel strainer pressure gauge.
4— Fuel temperature indicator.

6779. Which of the following would give the first positive indication that a change-over from one fuel tank to another is needed?

1— Fuel pressure warning.
2— Fuel pressure gauge.
3— Fuel flowmeter.
4— Fuel quantity indicator.

6780. A fuel pressure warning switch contacts close and warning light is turned on when

1— a measured quantity of fuel has passed through it.
2— the fuel pressure system is saturated with entrained water.
3— the fuel flow stops.
4— the fuel pressure drops below specified limits.

6781. A transmitter in a fuel pressure warning system serves what function?

1— Transmits an electrical signal to fluid pressure.
2— Converts fluid pressure to an electrical signal.
3— Converts electrical power to fluid pressure.
4— Transmits fluid pressure directly to the indicator.

6782. Where is fuel pressure taken for the pressure warning signal on most aircraft engines?

1— Outlet side of the boost pump.
2— Fuel pressure line of the carburetor.
3— Between the fuel pump and the strainer.
4— Upstream of the fuel shutoff valve.

6783. Which of the following is necessary to effectively troubleshoot a fuel pressure warning system?

1— The manufacturer's parts manual.
2— The manufacturer's maintenance manuals.
3— AC 43.13-1A, Acceptable Methods, Techniques, and Practices — Aircraft Inspection and Repair.
4— A set of Federal Aviation Regulations.

6784. Which of the following would be most useful to locate and troubleshoot an internal fuel leak in a large aircraft fuel system?

1— AC 43.13-1A, Acceptable Methods, Techniques, and Practices — Aircraft Inspection and Repair.
2— Federal Aviation Regulations.
3— A fuel system schematic.
4— AC 43.13-2A, Acceptable Methods, Techniques, and Practices — Aircraft Alterations.

6785. In some aircraft with several fuel tanks, the possible danger of allowing the fuel supply in one tank to become exhausted before the selector valve is switched to another tank is prevented by the installation of

1— a fuel pressure warning signal system.
2— a fuel pressure relief valve.
3— an engine fuel pump bypass valve.
4— a fuel flowmeter bypass valve.

6786. (1) The function of a fuel heater is to protect the engine fuel system from ice formation.

(2) An aircraft fuel heater cannot be used to thaw ice in the fuel screen.

Regarding the above statements,

1— only No. 1 is true.
2— only No. 2 is true.
3— both No. 1 and No. 2 are true.
4— neither No. 1 nor No. 2 is true.

6787. (1) Gas-turbine-engine fuel systems are very susceptible to the formation of ice in the fuel filters.

(2) A fuel heater operates as a heat exchanger to warm the fuel.

Regarding the above statements,

1—only No. 1 is true.
2—only No. 2 is true.
3—both No. 1 and No. 2 are true.
4—neither No. 1 nor No. 2 is true.

6788. (1) A fuel heater can use engine bleed air as a source of heat.

(2) A fuel heater can use engine lubricating oil as a source of heat.

Regarding the above statements,

1—only No. 1 is true.
2—only No. 2 is true.
3—both No. 1 and No. 2 are true.
4—neither No. 1 nor No. 2 is true.

6789. (1) A fuel pressure gauge is a differential pressure indicator.

(2) A fuel pressure gauge indicates the pressure of the fuel entering the carburetor.

Regarding the above statements,

1—only No. 1 is true.
2—only No. 2 is true.
3—both No. 1 and No. 2 are true.
4—neither No. 1 nor No. 2 is true.

6790. (1) A fuel pressure relief valve is required on an aircraft positive-displacement fuel pump.

(2) A fuel pressure relief valve is required on an aircraft centrifugal fuel boost pump.

Regarding the above statements,

1—only No. 1 is true.
2—only No. 2 is true.
3—both No. 1 and No. 2 are true.
4—neither No. 1 nor No. 2 is true.

6791. The primary purpose of a fuel tank sump is to provide a

1—positive system of maintaining the design minimum fuel supply for safe operation.
2—means of visually checking the amount of fuel actually in the tank prior to flight.
3—place where water and dirt accumulations in the tank can collect and be drained.
4—reserve supply of fuel to enable the aircraft to land safely in the event of fuel exhaustion.

6792. Why are integral fuel tanks used in many large aircraft?

1—To reduce fire hazards.
2—To facilitate servicing.
3—To minimize leakage.
4—To reduce weight.

6793. If an aircraft is fueled from a truck or storage tank which is known to be uncontaminated with dirt or water, periodic checks of the aircraft's fuel tank sumps and system strainers

1—need not be made except on aircraft carrying passengers for hire or used to give student instruction.
2—can be eliminated except for the strainer check before the first flight of the day and the fuel tank sump check during 100-hour or annual inspections.
3—are still necessary due to the possibility of contamination from other sources.
4—can be sharply reduced since contamination from other sources is relatively unlikely and of little consequence in modern aircraft fuel systems.

6794. Aircraft defueling should be accomplished

1—as soon as possible after engine shutdown.
2—with the aircraft's communication equipment on and in contact with the tower in case of fire.
3—in a hangar where activities can be controlled.
4—in the open air for good ventilation.

6795. Integral fuel tanks are

1—constructed of nonmetallic material.
2—readily removed from the aircraft.
3—located within the cabin space of an aircraft.
4—formed by the aircraft structure.

6796. What precautions must be observed if a gravity-feed fuel system is permitted to supply fuel to an engine from more than one tank at a time?

1—The tank airspaces must be interconnected.
2—The fuel outlet ports of each tank must have the same cross-sectional area.
3—Each tank must have a valve in its outlet that automatically shuts off the line when the tank is empty.
4—The tanks must always be serviced with identical amounts of fuel.

6797. The purpose of the baffle plate in a fuel tank is to

1—provide an expansion space for the fuel.
2—resist fuel surging within the fuel tank.
3—provide internal structural integrity.
4—prevent fuel overflow during refueling.

6798. What minimum required markings must be placed on or near each appropriate fuel filler cover on transport category aircraft?

1—The word "fuel," the minimum fuel grade or designation for the engines, and the total fuel tank capacity.
2—The minimum fuel grade and the total fuel tank capacity.
3—The word "fuel," the minimum fuel grade or designation for the engines, and the usable fuel tank capacity.
4—The word "fuel," the minimum fuel grade or permissible fuel designations, and the maximum permissible fueling and defueling pressure for pressure fueling systems.

6799. What is one disadvantage of using aromatic aviation fuels?

1—A fuel intercooler is required.
2—Deteriorates rubber parts.
3—Causes excessively rich mixture.
4—Results in low fuel volatility.

6800. Fuel-boost pumps are operated

1—to provide a positive flow of fuel to the engine.
2—during takeoff only.
3—primarily for fuel transfer.
4—automatically from fuel pressure.

6801. Flapper valves are used in fuel tanks to

1—reduce pressure.
2—prevent a negative pressure.
3—serve as variable restrictors.
4—act as check valves.

6802. Why are centrifugal-type boost pumps used in fuel systems of aircraft operating at high altitude?

1—To save space by having pumps submerged in the fuel tank.
2—Because they are positive displacement pumps.
3—To supply fuel under pressure to engine-driven pumps.
4—To permit air to circulate around the motor.

6803. Why is it necessary to vent all aircraft fuel tanks?

1—To ensure a positive head pressure for a submerged boost pump.
2—To exhaust fuel vapors.
3—To limit pressure differential between the tank and atmosphere.
4—To eliminate vapor locks.

6804. What minimum required markings must be placed on or near each appropriate fuel filler cover on small airplanes?

1—The word "fuel" and the minimum fuel grade or designation for the engine.
2—The word "fuel" and the total fuel capacity.
3—The word "fuel" and usable fuel capacity.
4—The word "fuel."

6805. The location of leaks and defects within the internal portions of the fuel system can usually be determined by

1—draining the system and purging with regulated air pressure.
2—visualizing the path of flow of the fuel from the fuel tank to the fuel metering device.
3—performing a fuel flow check.
4—observing the pressure gauge and operating the selector valves.

6806. What type of fuel-booster pump requires a pressure-relief valve?

1—Wobble.
2—Concentric.
3—Sliding vane.
4—Centrifugal.

6807. To prevent vapor lock in fuel lines at high altitude, some aircraft are equipped with

1—vapor separators.
2—direct-injection-type carburetors.
3—booster pumps.
4—vapor resistors.

6808. A fuel temperature indicator is located in the fuel tanks on some turbine-powered airplanes to tell when the fuel may be

1— getting cold enough to form hard ice.
2— in danger of forming ice crystals.
3— getting too cold to burn.
4— about to form rime ice.

6809. When inspecting a fuel system, you should check all valves located downstream of boost pumps with the pumps

1— removed from the structure.
2— at idle.
3— dormant.
4— operating.

6810. The type of fuel-boost pump that separates air and vapor from the fuel before it enters the line to the carburetor is the

1— diaphragm-type pump.
2— gear-type pump.
3— centrifugal-type pump.
4— sliding vane-type pump.

6811. Pressure fueling systems vary with aircraft manufacturers; therefore,

(1) the fueling operator should consult the aircraft manufacturer's instructions for detailed procedures.

(2) it is not necessary to consult the aircraft manufacturer's instructions when defueling a gravity fueling system.

Regarding the above statements,

1— only No. 1 is true.
2— only No. 2 is true.
3— both No. 1 and No. 2 are true.
4— neither No. 1 nor No. 2 is true.

6812. (1) Most aircraft pressure fueling systems consist of a pressure fueling hose and a panel of controls and gauges that permit one person to fuel or defuel any or all fuel tanks of an aircraft.

(2) Because of the limited fuel tank area, there are more advantages of a pressure fueling system in light aircraft.

Regarding the above statements,

1— only No. 1 is true.
2— only No. 2 is true.
3— both No. 1 and No. 2 are true.
4— neither No. 1 nor No. 2 is true.

6813. Which of the following could result if aviation gasoline does not vaporize readily enough?

A. Hard starting.
B. Poor acceleration.
C. Vapor locking.
D. Excessive gum deposits.
E. High cylinder temperatures.
F. Uneven fuel distribution.
G. Excessive crankcase dilution.

1— A, B, F, G.
2— B, C, D, E.
3— C, D, E, F.
4— D, E, F, G.

6814. (1) If aviation gasoline vaporizes too readily, fuel lines may become filled with vapor and cause increased fuel flow.

(2) A measure of a gasoline's tendency to vapor lock is obtained from the Reid vapor pressure test.

Regarding the above statements,

1— only No. 1 is true.
2— only No. 2 is true.
3— both No. 1 and No. 2 are true.
4— neither No. 1 nor No. 2 is true.

6815. Microbial growth is produced by various forms of micro-organisms that live and multiply in the water interfaces of jet fuels. Which of the following could result if microbial growth exists in a jet fuel tank and is not corrected?

A. Interference with fuel flow.
B. Interference with fuel quantity indicators.
C. Engine seizure.
D. Electrolytic corrosive action in a metal tank.
E. Lower grade rating of the fuel.
F. Electrolytic corrosive action in a rubber tank.

1— A, B, D.
2— B, C, E.
3— C, D, F.
4— A, E, F.

6816. Contamination is more of a problem in turbine fuel than it is in aviation gasoline because turbine fuel is

1— more viscous than gasoline.
2— less viscous than gasoline.
3— more plasticus than gasoline.
4— less plasticus than gasoline.

6817. The vapor pressure of aviation gasoline is

1— lower than the vapor pressure of automotive gasoline.
2— higher than the vapor pressure of automotive gasoline.
3— the same as the vapor pressure of automotive gasoline.
4— approximately 20 pounds per square inch at 100 °F.

6818. What can be done to eliminate or minimize the microbial growth problem in an aircraft jet fuel tank?

1— Use anti-icing and antibacterial additives.
2— Add CO_2 as a purgative.
3— Flush fuel tank regularly with a 5 percent solution of nitric acid.
4— Keep the fuel tank topped off.

6819. What is the maximum vapor pressure allowable for an aircraft fuel?

1— 7 PSI.
2— 5 PSI.
3— 3 PSI.
4— 9 PSI.

6820. If a bladder-type fuel tank is to be left empty for an extended period of time, the inside of the tank should be coated with a film of

1— engine oil.
2— methyl ethyl ketone.
3— linseed oil.
4— ethylene glycol.

6821. How may the antiknock characteristics of a fuel be improved?

1— By adding a knock inhibitor.
2— By adding a biocidal agent.
3— By adding a fungicide agent.
4— By adding a glycerin base additive.

6822. Some electric motors have two sets of field windings wound in opposite directions so that the

1— speed of the motor can be more closely controlled.
2— power output of the motor can be more closely controlled.
3— motor can be operated at any speed within its rated range without a change in power output.
4— motor can be operated in either direction.

6823. The purpose of a growler test is to determine the presence of

1— an out-of-round commutator.
2— a broken field lead.
3— a shorted armature.
4— a short from F+ to A-.

6824. If a short circuit occurs between the positive armature lead and the field lead of a shunt generator which has the voltage regulator located in the positive side of the field circuit, the

1— generator voltage will drop to zero.
2— generator will only produce residual voltage.
3— reverse current cutout relay will open and remain open until the fault is corrected.
4— generator voltage will increase.

6825. What is the principal advantage of the series-wound dc motor?

1— High starting torque.
2— Suitable for constant speed use.
3— Low starting torque.
4— Speed slightly higher when unloaded.

6826. If a generator is equipped with a vibrator-type voltage regulator, the actual time the voltage regulator points remain open

1— depends on the load carried by the generator.
2— is controlled by the current limiter point clearance.
3— is controlled by the reverse-current cutout relay point clearance.
4— is increased when the external load is greater than the generator output.

6827. What is a cause of generator brush arcing?

1— Seating brushes with No. 000 sandpaper.
2— Excessive spring tension.
3— Carbon dust particles.
4— Low spring tension.

6828. When ac generators are operated in parallel, the

1— amperes and frequency must both be equal.
2— wattage and voltage must both be equal.
3— frequency and voltage must both be equal.
4— amperes and voltage must both be equal.

6829. The starting current of a series-wound dc motor, in passing through both the field and armature windings, produces a

1— low starting torque.
2— speed slightly higher when unloaded.
3— high starting torque.
4— force for constant speed.

6830. Which motor would be most likely to have an armature brake?

1— Starter motor.
2— Landing light retraction motor.
3— Inverter drive motor.
4— Anticollision beacon operating motor.

6831. The method most often used in overcoming the effect of armature reaction is through the use of

1— interpoles.
2— ALNICO field pieces.
3— shaded poles.
4— drum-wound armatures in combination with a negatively connected series field.

6832. The only practical method of maintaining a constant voltage output from an aircraft generator under varying conditions of speed and load is to vary the

1— strength of the magnetic field.
2— number of conductors in the armature.
3— speed at which the armature rotates.
4— brush pressure on the commutator segments.

6833. The pole pieces or shoes used in a dc generator are a part of the

1— commutator assembly.
2— armature assembly.
3— field assembly.
4— brush assembly.

6834. How many cycles of ac voltage are produced in a six-pole alternator of the revolving-field type for each revolution of the rotor?

1— Four.
2— Five.
3— Three.
4— Six.

6835. If the reverse current cutout relay contact points fail to open after the generator output has dropped below battery potential, current will flow through the generator armature

1— in the normal direction and through the shunt field opposite the normal direction.
2— and the shunt field opposite the normal direction.
3— and the shunt field in the normal direction.
4— opposite the normal direction and through the shunt field in the normal direction.

6836. How does the magnetic brake used to stop rotation of an electric motor armature operate?

1— Centrifugal force releases a rotating brake cog from a stationary notch when the armature reaches a certain speed and magnetic force re-engages the cog when the electrical power is turned off.
2— A friction brake is applied by a magnet and released by a spring.
3— A friction brake is applied by a spring and released by a magnet.
4— A brake winding is installed in the rotating armature to cause a more rapid collapse of the magnetic flux lines when the electric power is turned off.

6837. In a generator, what eliminates any possible sparking to the brush guides caused by the movement of the brushes within the holder?

1— The brush pigtail.
2— Brush spring tension.
3— Undercutting the mica on the commutator.
4— Lubricating the brush sides.

6838. A series-wound dc electric motor will normally require

1— more current at high RPM than at low RPM.
2— approximately the same current throughout its operating range of speed.
3— more current at low RPM than at high RPM.
4— none of the above because a series-wound motor is practical for operation only with alternating current.

6839. The effect of armature reaction (the amount that the neutral plane shifts) is proportional to the

1— strength of the stationary field.
2— voltage output of the generator.
3— speed (RPM) of the generator.
4— load on the generator.

6840. Aluminum wire must be stripped very carefully because

1— high resistance will develop in stripping nicks.
2— low resistance will develop in stripping nicks.
3— stripping nicks will cause short circuits in wire runs.
4— individual strands will break easily after being nicked.

6841. The purpose of a commutator in a dc motor is to

1— allow the transfer of current by converting it.
2— reverse the current in the field coils at the proper time in order to maintain the current flow in the same direction in all conductors under a given pole.
3— provide a means of transferring mechanical energy.
4— reverse the current in the coil just at the time the coil becomes parallel to the lines of force.

6842. An ammeter in a battery charging system is for the purpose of indicating the

1— amperage available for use.
2— total amperes being used in the airplane.
3— rate of current used to charge the battery.
4— electrical potential of battery.

6843. Which of the following is not one of the purposes of interpoles in a generator?

1— Counteract field distortion.
2— Reduce field strength.
3— Overcome armature reaction.
4— Reduce arcing at the brushes.

6844. To test generator or motor armature windings for opens,

1— place armature in a growler and connect a 110V test light on adjacent segments; light should light.
2— check adjacent segments on commutator with an ohmmeter on the high resistance scale.
3— use a 12/24V test light between the armature core segments and the shaft.
4— use a 110V test light between the armature core segments and the shaft.

6845. The nominal current rating of an aircraft switch is normally stamped on the

1— switch housing.
2— face plate.
3— inside of the switch.
4— nonmetallic housing.

6846. Where are position lights for navigation located on all civil airplanes?

1— Aft (L & R) and forward.
2— Forward (L & R) and aft.
3— Upper (L & R) and lower (L & R).
4— Upper, lower, forward, and aft.

6847. Which of the following establishes the minimum standards concerning the number, color, location, and intensity of position lights for navigation on civil aircraft?

1— Aircraft Owners and Pilots Association.
2— Federal Aviation Administration.
3— Air Transport Association.
4— Aircraft manufacturers.

6848. To what depth is the mica insulation between the commutator bars of a dc generator undercut?

1— One-half the width of the mica.
2— Equal to twice the width of the mica.
3— Equal to the width of the mica.
4— Never undercut.

6849. A voltage regulator controls generator output by

1— introducing a resistance in generator-to-battery lead in the event of overload.
2— shorting out field coil in the event of overload.
3— varying current flow to generator field coil.
4— motorizing generator to oppose its action.

6850. Aircraft operating at night must be equipped with position lights that meet the minimum requirements specified by the

1— safety control regulations.
2— air safety regulations.
3— Federal Aviation Regulations.
4— air transportation board.

6851. Which type of dc generator is not used as an airplane generator?

1— Externally grounded.
2— Internally grounded.
3— Series wound.
4— Compound wound.

6852. What type ammeter is used to measure radio frequency alternating current?

1— Half-wave bridge type.
2— Full-wave bridge type.
3— Thermocouple type.
4— Emitter base type.

6853. What is the most accurate type of frequency-measuring instrument?

1— Integrated circuit chip having a clock circuit.
2— Electrodynamometers using electromagnetic fields.
3— Electromagnets using one permanent magnet.
4— Repulsion-type, moving-vane meter.

6854. During ground operation, a starter-generator is normally cooled by

1— ram air.
2— engine bleed air.
3— an integral fan.
4— the environmental system cooled air.

6855. During flight operation, a starter-generator is cooled by

1— the environmental system cooled air.
2— an integral fan and ram air.
3— engine bleed air.
4— an external motor-driven fan.

6856. What does a rectifier do?

1— Changes direct current into alternating current.
2— Steps up voltage.
3— Changes alternating current into direct current.
4— Reduces voltage.

6857. What type of instrument is used for measuring very high values of resistance?

1— Megohmmeter.
2— Shunt-type ohmmeter.
3— Thermocouple.
4— Multimeter.

6858. When a diode is checked for an open circuit or a short circuit, it should be

1— in the circuit.
2— checked with a milliamp ammeter.
3— disconnected from the circuit.
4— checked from positive to negative only.

6859. Why must caution be observed when handling a high-voltage capacitor in an electrical circuit?

1— A capacitor may emit toxic gases if not properly ventilated.
2— The polarity of the plates may be reversed by improper attachment of an ohmmeter.
3— A capacitor may lose its ability to hold charge if intentionally discharged.
4— A capacitor may retain its charge after power is removed.

6860. What type of capacitor can be checked with an ohmmeter to determine its condition?

1— An extremely low capacity capacitor.
2— A sodium-filled capacitor.
3— An electrolytic capacitor.
4— A dielectric capacitor.

6861. How can it be determined if a transformer winding has some of its turns shorted together?

1— Measure the input voltage with an ohmmeter.
2— The output voltage will be high.
3— The transformer will not function.
4— The transformer will get hot in normal operation.

6862. Which of the following are the major parts of a dc motor?

A. Armature assembly.
B. Field assembly.
C. Brush assembly.
D. Commutator.
E. Pole piece.
F. Rheostat.
G. End frame.

1— A, B, C, G.
2— B, C, D, E.
3— A, D, F, G.
4— C, E, F, G.

6863. (1) There are three basic types of dc motors; series, shunt, and compound.

(2) In the series motor, the field windings, consisting of relatively few turns of heavy wire, are connected in series with the armature winding.

Regarding the above statements,

1— only No. 1 is true.
2— only No. 2 is true.
3— both No. 1 and No. 2 are true.
4— neither No. 1 nor No. 2 is true.

6864. What are the two general types of ac motors used in aircraft systems?

1— Induction and synchronous.
2— Shaded pole and universal.
3— AC series and capacitor start.
4— Rheostat series and condenser start.

6865. Which of the following factors must be taken into consideration when determining the wire size to use for an aircraft installation?

A. Mechanical strength.
B. Allowable power loss.
C. Ease of installation.
D. Resistance of current return path through the aircraft structure.
E. Permissible voltage drop.
F. Current carrying capability of the conductor.
G. Type of load (continuous or intermittent).

1—B, E, F, G.
2—A, B, D, E.
3—C, D, F, G.
4—A, C, D, F.

6866. When selecting hardware for attaching bonding connections to an aircraft structure, which of the following should be considered?

A. Mechanical strength.
B. Allowable power loss.
C. Ease of installation.
D. Permissible voltage drop.
E. Amount of current to be carried.
F. Type of load (continuous or intermittent).

1—A, C, E.
2—B, D, F.
3—D, E, F.
4—A, B, C.

6867. How should the splices be arranged if several are to be located in an electrical wire bundle?

1—Staggered along the length of the bundle.
2—Grouped together.
3—Enclosed in a conduit.
4—It is not permissible to splice aircraft wiring.

6868. What is the minimum bend radii for an electrical wire bundle?

1—Ten times the outside diameter of the bundle.
2—Five times the outside diameter of the bundle.
3—Fifteen times the outside diameter of the bundle.
4—Twenty times the diameter of the largest wire in the bundle.

6869. The voltage output of an alternator may be regulated by controlling the

1—speed of the alternator.
2—voltage output of the dc exciter.
3—resistance in the rotor windings.
4—exciter frequency.

6870. If several long lengths of electrical cable are to be installed in rigid conduit, the possibility of damage to the cable as it is pulled through the conduit will be reduced by

1—dusting the cable with powdered graphite.
2—dusting the cable with powdered soapstone.
3—blowing powdered graphite into the conduit runs before installation of the cables.
4—applying a light coat of oil or grease.

6871. Grounding is electrically connecting a conductive object to the primary structure in order to

1—prevent current return paths.
2—prevent stability of radio transmission and reception.
3—allow accumulation of static charge.
4—prevent development of radio frequency potential.

6872. What is normally used to bond noncontinuous stainless steel aircraft components?

1—Printed circuits.
2—Stainless steel jumpers.
3—Copper jumpers.
4—Aluminum jumpers.

6873. Aircraft fuse capacity is rated in

1—volts.
2—ohms.
3—amperes.
4—microfarads.

6874. When adding a rheostat to a light circuit to control the light intensity, it should be connected

1—in parallel with the light.
2—across the source of energy.
3—in series with the light.
4—in series parallel with the light switch.

6875. What will prevent the inadvertent operation of a switch?

1—Mount a suitable guard over the switch.
2—Install a derated spring-loaded toggle switch.
3—Install circuit breakers as a switch.
4—Install a low amperage fuse across the contacts.

6876. If one switch is used to control all navigation lights, the lights are most likely connected

1— in series with each other and parallel to the switch.
2— in series with each other and in series with the switch.
3— parallel to each other and parallel to the switch.
4— parallel to each other and in series with the switch.

6877. Why is it important to prevent oil canning in electrical junction box sides?

1— To aid in the installation of terminal strips.
2— To prevent internal short circuits.
3— To provide space for relay switches.
4— To provide space for wire ties and clamps.

6878. Electric wiring installed in aircraft without special enclosing means (open wiring) offers the advantages of ease of installation, simple maintenance, and reduced weight. When bundling open wiring, the bundles should

1— be limited as to the number of cables to minimize damage from a single electrical fault.
2— include at least one shielded cable to provide good bonding of the bundle to the airframe.
3— be limited to a minimum bend radius of five times the bundle diameter to avoid excessive stresses on the cable insulation.
4— be covered with soft insulating tubing (spaghetti) for mechanical protection where severe external abrasion is likely to occur.

6879. During inspection of the terminal strips of an aircraft electrical system, it should be determined that

1— only locknuts have been used for terminal attachment to the studs.
2— the terminal studs are anchored against rotation.
3— the heat rise at any single terminal stud does not exceed 120 °F above ambient temperatures.
4— only plain nuts and lockwashers have been used for terminal attachment to the studs.

6880. What protection to wires and cables does conduit provide when used in aircraft installations?

1— Electromagnetic.
2— Thermal.
3— Mechanical.
4— Structural.

6881. Which of the following should be avoided in conduit installation?

1— Support the conduit to prevent chafing against the structure.
2— Provide drainholes at the lowest point in a conduit run.
3— Locate conduit to provide a footstep or handhold for personnel.
4— Drilling burrs should be carefully removed.

6882. If the (+) terminal of a voltmeter is connected to the (-) terminal of the source voltage and the (-) terminal of the meter is connected to the (+) terminal of the source voltage, the voltmeter will read

1— high voltage.
2— correctly.
3— low voltage.
4— backwards.

6883. When using the voltage drop method of checking circuit resistance, the

1— output voltage must be maintained at an intermittent value.
2— input voltage must be maintained at a constant value.
3— output voltage must be maintained at a constant value.
4— input voltage must be varied.

6884. The nominal rating of electrical switches refers to continuous

1— voltage rating with the contacts open.
2— current rating with the contacts open.
3— voltage rating with the contacts closed.
4— current rating with the contacts closed.

6885. Aircraft electrical junction boxes located in a fire zone are usually constructed of

1— fireproof aluminum.
2— asbestos.
3— cadmium-plated steel.
4— stainless steel.

6886. Which conduit is normally used where necessary to minimize radio interference?

1— Flexible brass.
2— Flexible aluminum.
3— Rigid steel.
4— Rigid aluminum.

6887. The primary considerations when selecting electric cable size are

1— current-carrying capacity and allowable voltage drop.
2— the voltage and amperage of the load it must carry.
3— the cable location and normal operating temperature.
4— the system voltage and cable length.

6888. The navigation lights of some aircraft consist of a single circuit controlled by a single switch which has an ON position and an OFF position, with no additional positions possible. This switch is referred to as a

1— double-pole, single-throw (DPST), two-position switch.
2— single-pole, double-throw (SPDT), two-position switch.
3— double-pole, double-throw (DPDT), two-position switch.
4— single-pole, single-throw (SPST), two-position switch.

6889. Electric circuits are protected from overheating by means of

1— thermocouples.
2— shunts.
3— fuses.
4— solenoids.

6890. How does the routing of coaxial cables differ from the routing of electrical wiring?

1— Coaxial cables are routed parallel with stringers or ribs.
2— Coaxial cables are routed at right angles to stringers or ribs.
3— Coaxial cables must not be clamped.
4— Coaxial cables are routed as directly as possible.

6891. Which of the following aluminum electrical cable sizes should be selected to replace a No. 10 copper electrical cable?

1— No. 4.
2— No. 6.
3— No. 8.
4— No. 10.

6892. In installations where the ammeter is in the generator or alternator lead, and the regulator system does not limit the maximum current that the generator or alternator can deliver, the ammeter can be redlined at

1— 50 percent of the generator or alternator rating.
2— 60 percent of the generator or alternator rating.
3— 75 percent of the generator or alternator rating.
4— 100 percent of the generator or alternator rating.

6893. Which statement relating to electric wiring is true?

1— When attaching a terminal to the end of an electric cable, it should be determined that the strength of the cable-to-terminal joint is at least twice the tensile strength of the cable.
2— When splicing an electric cable in a location subjected to extreme vibration, it is generally recommended that solder splices be used.
3— When attaching a terminal to the end of an electric cable, it should be determined that the strength of the cable-to-terminal joint is at least equal to the tensile strength of the cable itself.
4— All electric cable splices should be covered with soft insulating tubing (spaghetti) for mechanical protection against external abrasion.

6894. Bonding connections should be tested for

1— resistance value.
2— amperage value.
3— reactance.
4— voltage value.

6895. What kind of switch should you install in a single wire circuit that required the switch to be manually held in the ON position?

1— Single-pole, single-throw (SPST), two-position normally open (NO).
2— Single-pole, single-throw (SPST), single-position.
3— Single-pole, double-throw (SPDT), two-position.
4— Single-pole, double-throw (SPDT), single-position normally open (NO).

6896. A circuit breaker is installed in an aircraft electrical system primarily to protect the

1— circuit and should be located as close to the source as possible.
2— circuit and should be located as close to the unit as possible.
3— electrical unit in the circuit and should be located on the unit side of the switch.
4— electrical unit in the circuit and should be located as close to the source as possible.

6897. How should a voltmeter be connected?

1— In series with the source.
2— In parallel with the load.
3— In series with the load.
4— In series parallel with the source.

6898. If aluminum electrical cable is to be substituted for No. 12 copper electrical cable, which gauge should be selected?

1— No. 4.
2— No. 8.
3— No. 6.
4— No. 10.

6899. If it is necessary to use an electrical connector where it may be exposed to moisture, the mechanic should

1— coat the connector with grease.
2— use a special moisture-proof type.
3— wrap the connector with waxed paper.
4— spray the connector with varnish or zinc-chromate.

6900. The two kinds of fuse holders used most commonly in aircraft circuits are

1— clip-on, screw-in types.
2— plug-in, clip types.
3— plug-in, mechanical reset types.
4— mechanical reset, screw-in types.

6901. If a wire is installed so that it comes in contact with some moving parts, what protection should be given the wire?

1— Wrap with soft wire solder into a shield.
2— Wrap with friction tape.
3— Pass through conduit.
4— Several coats of Lionoil or varnish.

6902. In the American Wire Gauge (AWG) system of numbers used to designate electrical wire sizes, the number assigned to a size is related to its

1— combined resistance and current-carrying capacity.
2— current-carrying capacity.
3— approximate electrical resistance per 1,000 feet.
4— cross-sectional area.

6903. Why are aircraft components bonded?

1— To allow electrical charges to move through the aircraft structure without causing sparks.
2— To prevent electrical charges from moving through the aircraft structure.
3— To maintain the electrostatic charge of the aircraft equal to that of the surrounding atmosphere.
4— To allow the electrostatic charge of the aircraft to dissipate before it contacts the ground after flight.

6904. What is the allowable voltage drop for a No. 18 copper wire 50 feet long to carry 12.5 amperes, continuous operation?

Use the formula VD = RLA

VD = Voltage drop
R = Resistance per ft. = .00644
L = Length of wire
A = Amperes

1— 1/2V.
2— 7V.
3— 1V.
4— 4V.

6905. What is the purpose of the selection of derated switches for known continuous load current applications?

1— To calculate the voltage drop across the circuit.
2— To prevent contact oscillation in high rush-in circuits.
3— To prevent short circuits in the motor field windings.
4— To obtain reasonable switch efficiency and service life.

6906. What is the advantage of a circuit breaker when compared to a fuse?

1— Never needs replacing.
2— Responds faster to overload.
3— Always eliminates the need of a switch.
4— Resettable and reusable.

6907. What is the advantage of a current limiter?

1— It breaks circuit quickly.
2— It can be reset easily.
3— It is easily replaced.
4— It will take overload for a short period.

6908. Where electric cables must pass through holes in bulkheads, formers, ribs, firewalls, etc., the wires should be protected from chafing by

1— wrapping with tape.
2— using a rubber grommet.
3— several coats of varnish.
4— wrapping with plastic.

6909. In aircraft electrical systems, automatic reset circuit breakers

1— should not be used as circuit protective devices.
2— are useful where only temporary overloads are normally encountered.
3— need not be made accessible to crewmembers in flight.
4— must be used in all circuits essential to safe operation of the aircraft.

6910. A certain switch is described as a single-pole, double-throw switch (SPDT). The throw of a switch indicates

1— the number of circuits each pole can complete through the switch.
2— the method of actuating the switch (push-pull, laterally, vertically, etc.).
3— the number of terminals at which current can enter or leave the switch.
4— the number of places at which the operating device (toggle, plunger, etc.) will come to rest and at the same time open or close a circuit.

6911. When considering an alteration, the criteria upon which the selection of electric cable size should be based are

1— applied voltage and wire length.
2— applied voltage and allowable voltage drop.
3— current-carrying capacity and allowable voltage drop.
4— current-carrying capacity and applied voltage.

6912. What is an important factor in selecting aircraft fuses?

1— The current exceeds a predetermined value.
2— The voltage rating should be lower than the maximum circuit voltage.
3— The inner strip of metal is made of an alloy of tin and bismuth.
4— Capacity matches the needs of the circuit.

6913. The circuit breaker in the instrument lighting system protects the

1— lights from too much current.
2— wiring from too much current.
3— wiring from too much voltage.
4— lights from too much voltage.

6914. One advantage of using ac electrical power in aircraft is

1— that ac electrical motors can be reversed while dc motors cannot.
2— that self-induction due to voltage change contributes to the effective power, thus causing the power output to be 1.707 times the power input.
3— greater ease in stepping the voltage up or down.
4— that the effective voltage is 1.41 times the maximum instantaneous voltage; therefore, less power input is required.

6915. Aircraft position lights consist of at least three lights. Their color and location are

1— white in front, red in the rear, and green midway on the aircraft centerline.
2— red on the left, green on the right, and white on the rear.
3— green in front, red in the rear, and white midway on the aircraft centerline.
4— red on the right, green on the left, and white in the rear.

6916. Why are the iron cores of most induction coils laminated?

1— To reduce the core reluctance.
2— To increase the core permeability.
3— To reduce the effects of eddy currents.
4— To reduce the production of weak areas and strong areas on the core faces.

6917. Certain transport aircraft use ac electrical power for all normal operation and battery furnished dc electrical power for standby emergency use. In aircraft of this type that operate no dc generators, the batteries are kept charged by

1— inverters which use the aircraft's ac generators as a source of power.
2— ac current directly from the aircraft's generators.
3— alternators which use the aircraft's generators as a source of power.
4— rectifiers which use the aircraft's ac generators as a source of power.

6918. The voltage in an ac transformer secondary that contains twice as many loops as the primary will be

1— greater and the amperage less than in the primary.
2— greater and the amperage greater than in the primary.
3— less and the amperage greater than in the primary.
4— less and the amperage less than in the primary.

6919. If the positive field lead between a generator and a generator control panel breaks and is shorted while the engine is running, a voltmeter connected to generator output would indicate

1— zero voltage.
2— residual voltage.
3— normal voltage.
4— slightly below normal voltage.

6920. What is a method used for restoring generator field residual magnetism?

1— Flash the fields.
2— Demagnetize the commutator.
3— Reseat the brushes.
4— Energize the armature.

6921. One of the chief advantages of alternating current is that it can be transmitted at a high voltage with a low power loss; the voltage can then be changed to any desired value of

1— dc by means of inverters.
2— dc by means of transformers.
3— ac by means of inverters.
4— ac by means of transformers.

6922. Which of the following must be accomplished when installing an anticollision light?

1— Connect the light to the primary electrical bus.
2— Install a switch independent of the position light switch.
3— Use shielded electrical cable to assure fail-safe operation.
4— Connect the anticollision light to the aircraft position light switch.

6923. The inductor-type inverter output voltage is controlled by the

1— number of poles and the speed of the motor.
2— voltage regulator.
3— dc stator field current.
4— ac armature coils.

6924. When using an ohmmeter to check the continuity of a generator field coil, the coil should

1— be removed from the generator housing.
2— show high resistance when the meter prods are connected to the terminals of the coil.
3— be heated to operating temperature.
4— show very low resistance if it is a series field coil.

6925. The strength of the core of an electromagnet depends upon the material from which it is constructed and which of the following?

1— The number of turns of wire in the coil and the applied voltage.
2— The size (cross section) of the wire and the amount of current (amperes) passing through the coil.
3— The number of turns of wire in the coil and the amount of current (amperes) passing through the coil.
4— The size (cross section) and the number of turns of wire in the coil and the applied voltage.

6926. A voltage regulator controls generator voltage by changing the

1— resistance in the generator output circuit.
2— residual magnetism of the generator.
3— current in the generator output circuit.
4— resistance of the generator field circuit.

6927. The overvoltage control automatically protects the generator system when excessive voltage is present by

1— opening the shunt field circuit.
2— opening and resetting the field control relay.
3— breaking a circuit to the trip coil of the field control relay.
4— closing the generator switch circuit.

6928. When dc generators are operated in parallel to supply power for a single load, their controls include an equalizer circuit to assure that all generators share the load equally. The equalizer circuit operates by

1— switching all new electrical loads to the low generator to maintain an equal load division among the generators.
2— increasing the output of the low generator to equal the output of the high generator.
3— decreasing the output of the high generator to equal the output of the low generator.
4— increasing the output of the low generator and decreasing the output of the high generator until they are equal.

6929. Which of the following is considered to be an intermittent duty circuit?

1— Anticollision light circuit.
2— Landing light circuit.
3— Instrument panel light circuit.
4— Navigation light circuit.

6930. The most common method of regulating the voltage output of a compound dc generator is to vary the

1— current flowing through the shunt field coils.
2— total effective field strength by changing the reluctance of the magnetic circuit.
3— resistance of the series field circuit.
4— number of rotating conductors being affected by the field flux.

6931. (Refer to figure 18.) Upon completion of the landing gear extension cycle, the green light illuminated and the red light remained lit. What is the probable cause?

1— Short in the down limit switch.
2— Short in the gear safety switch.
3— Short in the up limit switch.
4— Short in the nose gear down switch.

6932. If any one generator in a 24-volt dc system shows low voltage, the most likely cause is

1— the generator polarity is reversed.
2— an out-of-adjustment voltage regulator.
3— shorted or grounded wiring.
4— a defective reverse current cutout relay.

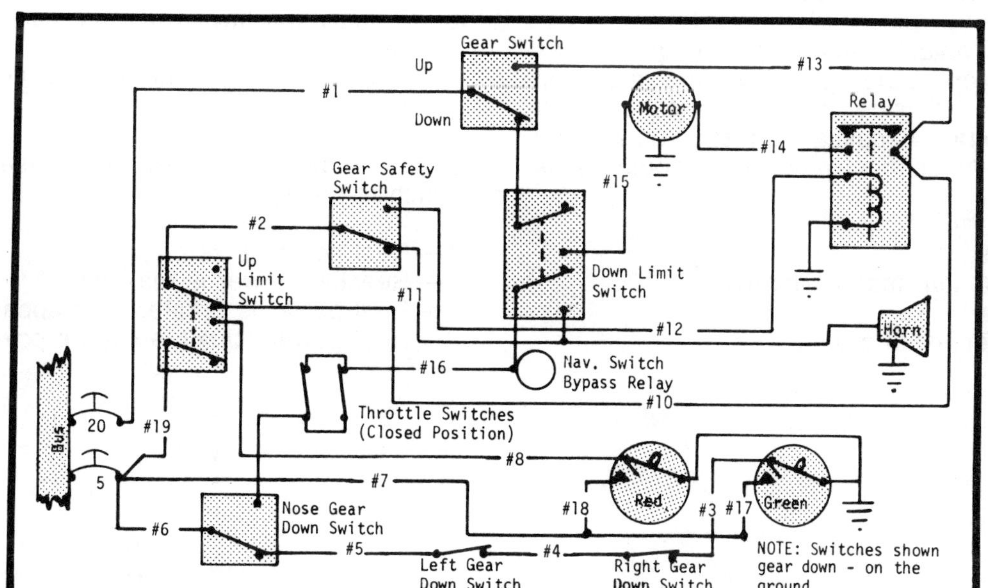

Figure 18

6933. How can the direction of rotation of a dc electric motor be changed?

1—Interchange the wires which connect the motor to the external power source.
2—Reverse the electrical connections to either the field or armature windings.
3—Rotate the brush assembly 90°.
4—Remove the starting winding.

6934. Aircraft which operate only ac generators (alternators) as a primary source of electrical power normally provide current suitable for battery charging through the use of

1—a stepdown transformer and a rectifier.
2—a network of condensers and choke coils to filter the alternating current with negligible power loss.
3—an inverter and a voltage-dropping resistor.
4—a dynamotor with a half-wave dc output.

6935. During inspection of an anticollision light installation for condition and proper operation, it should be determined that

1—electrical or mechanical interconnections are provided so that the anticollision light will operate at all times that the position light switch is in the ON position.
2—an appropriately rated fuse is in position at the light to protect the connecting wiring against electrical faults.
3—two independent circuits connect the light to the source of power so that the light will continue to operate in case of damage to one circuit.
4—the anticollision light can be operated independently of the position lights.

6936. Major adjustments on equipment such as regulators, contactors, and inverters are best accomplished outside the airplane on test benches with necessary instruments and equipment. Adjustment procedure should be as outlined by

1—textbooks.
2—the equipment manufacturer.
3—the FAA.
4—aircraft technical orders.

6937. A battery-generator system provides direct current. On installations requiring alternating current from the battery-generator system, it is necessary to have

1—a transformer.
2—two or more generators.
3—an inverter.
4—a variable resistor between the battery and generator.

6938. A relay is

1—a magnetically operated switch.
2—a device which increases voltage.
3—a device which converts electrical energy to heat energy.
4—any conductor which receives electrical energy and passes it on with little or no resistance.

6939. The purpose of a rectifier in an electrical system is to change

1—the frequency of alternating current.
2—the voltage of alternating current.
3—the voltage and amperage of alternating current.
4—alternating current to direct current.

6940. What is the ratio of turns between the primary coil winding and the secondary coil winding of a transformer designed to triple its input voltage?

1—Primary will have one-third as many turns as its secondary.
2—Primary will have one-half as many turns as its secondary.
3—Primary will have twice as many turns as its secondary.
4—Primary will have three times as many turns as its secondary.

6941. In an ac circuit with no phase lead or lag, which is true?

1—Real power is zero.
2—Reactive power is maximum.
3—Real power is greater than apparent power.
4—Real power equals apparent power.

6942. How are generators rated?

1— Watts at rated voltage.
2— Farads at rated voltage.
3— Amperes at rated voltage.
4— The impedance at rated voltage.

6943. How is a shunt-wound dc generator connected?

1— One field is shunted across the other.
2— Both fields are shunted across the armature.
3— The field and armature are shunted with a capacitor.
4— The armature and fields are shunted by a variable resistor.

6944. The poles of a generator are laminated to

1— reduce hysteresis losses.
2— reduce flux losses.
3— increase flux concentration.
4— reduce eddy current losses.

6945. What is the frequency of an alternator dependent upon?

1— Voltage.
2— RPM.
3— Current.
4— Wattage rating.

6946. The generator rating is usually found stamped on the

1— firewall.
2— generator.
3— engine.
4— cowling.

6947. Residual voltage is a result of

1— magnetism in the field windings.
2— current flow in the field coils.
3— magnetism in the field shoes.
4— magnetism in the armature.

6948. (Refer to figure 19.) Which of the batteries are connected together incorrectly?

1— A.
2— B.
3— C.
4— D.

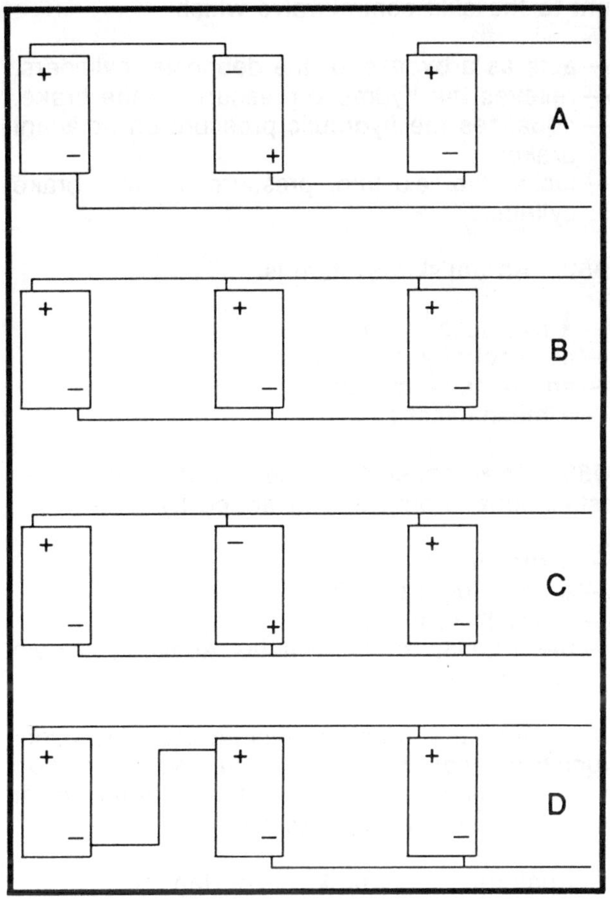

Figure 19

6949. In troubleshooting an electrical circuit, if an ohmmeter is properly connected across a circuit component and some value of resistance is read,

1— the component has continuity and is open.
2— either the component or the circuit is shorted.
3— the component has no continuity and is open.
4— the component has continuity and is not open.

6950. The purpose of antiskid generators is to

1— eliminate brake drag.
2— monitor hydraulic pressure applied to brakes.
3— indicate when a tire skid occurs.
4— measure wheel rotational speed and any speed changes.

6951. In a brake antiskid system, when an approaching skid is sensed, an electrical signal is sent to the skid control valve which

1— acts as a bypass for the debooster cylinders.
2— relieves the hydraulic pressure on the brake.
3— equalizes the hydraulic pressure on adjacent brakes.
4— locks the existing pressure in the brake cylinder.

6952. An antiskid system is

1— a hydraulic system.
2— an electrohydraulic system.
3— an electrical system.
4— a mechanical system.

6953. The antiskid brake system in most installations is activated or armed by

1— hydraulic pressure.
2— a centrifugal switch.
3— a switch in the cockpit.
4— the rotation of the wheels above a certain speed.

6954. An aural warning system provides audio signals whenever the throttle setting, in relation to the setting of which of the following flight controls, is unsafe for takeoff?

1— Spoilers, speed brake, and stabilizer.
2— Elevators, spoilers, and slats.
3— Speed brake, flaps, and stabilizer.
4— Spoilers, flaps, and elevators.

6955. The primary purpose of a takeoff warning system is to alert the crew that a flight control is not properly set prior to takeoff. It is actuated by the closing of the

1— landing gear retract switch.
2— ground proximity switch (squat switch).
3— thrust lever switch.
4— stabilizer switch.

6956. (1) An airspeed indicator measures the differential between pitot and static air pressures surrounding the aircraft at any moment of flight.

(2) An airspeed indicator measures the differential between pitot and cabin air pressures at any moment of flight and makes the indication mechanically.

Regarding the above statements,

1— neither No. 1 nor No. 2 is true.
2— both No. 1 and No. 2 are true.
3— only No. 2 is true.
4— only No. 1 is true.

6957. The angle-of-attack detector operates from differential pressure when the airstream

1— is parallel to the longitudinal axis of the aircraft.
2— is not parallel to the true angle of attack of the aircraft.
3— is parallel to the angle of attack of the aircraft.
4— is parallel to the angle of incidence.

6958. (1) When an airplane is slowed below approximately 20 miles per hour, the antiskid system automatically deactivates to give the pilot full control of the brakes for maneuvering and parking.

(2) An antiskid system consists basically of three components; wheel speed sensors, control box, and control valves.

Regarding the above statements,

1— only No. 1 is true.
2— only No. 2 is true.
3— both No. 1 and No. 2 are true.
4— neither No. 1 nor No. 2 is true.

6959. In an antiskid system, wheel skid is detected by

1— an electrical sensor.
2— a discriminator.
3— an annunciator.
4— a sudden rise in brake pressure.

6960. Which of the following functions does a skid control system perform?

A. Normal skid control.
B. Normal braking.
C. Fail safe protection.
D. Locked wheel skid control.
E. Touchdown protection.
F. Takeoff protection.

1—A, B, C, D.
2—A, C, D, E.
3—A, D, E, F.
4—A, B, E, F.

6961. What effect, if any, will a wheel going into a skid have on the aircraft's braking value?

1—Greatly reduced.
2—Slightly increased.
3—No effect.
4—Totally destroyed.

6962. At what point in the landing operation does normal skid control perform its function?

1—When wheel rotation has slowed but not come to a stop.
2—When wheel rotation has come to a stop.
3—When wheel rotation is at the maximum.
4—Anytime the wheel is rotating.

6963. (1) An antiskid system is designed to apply enough force to operate just below the skid point.

(2) A warning lamp lights in the cockpit when the antiskid system is turned off or if there is a system failure.

Regarding the above statements,

1—only No. 1 is true.
2—only No. 2 is true.
3—both No. 1 and No. 2 are true.
4—neither No. 1 nor No. 2 is true.

6964. (Refer to figure 20.) During a landing gear retraction test (aircraft on jacks) with one or more throttles retarded and the landing gear in any position between the retracted and down and locked position, the warning horn

1—blows and red position light extinguishes.
2—does not blow and the red position light extinguishes.
3—blows and red position light flashes.
4—blows and red position light illuminates.

6965. (Refer to figure 20). Which repair should be made if the gear switch was placed in UP position and the gear does not retract?

1—Replace electrical wire No. 11.
2—Replace electrical wire No. 15.
3—Replace the down limit switch.
4—Replace electrical wire No. 12.

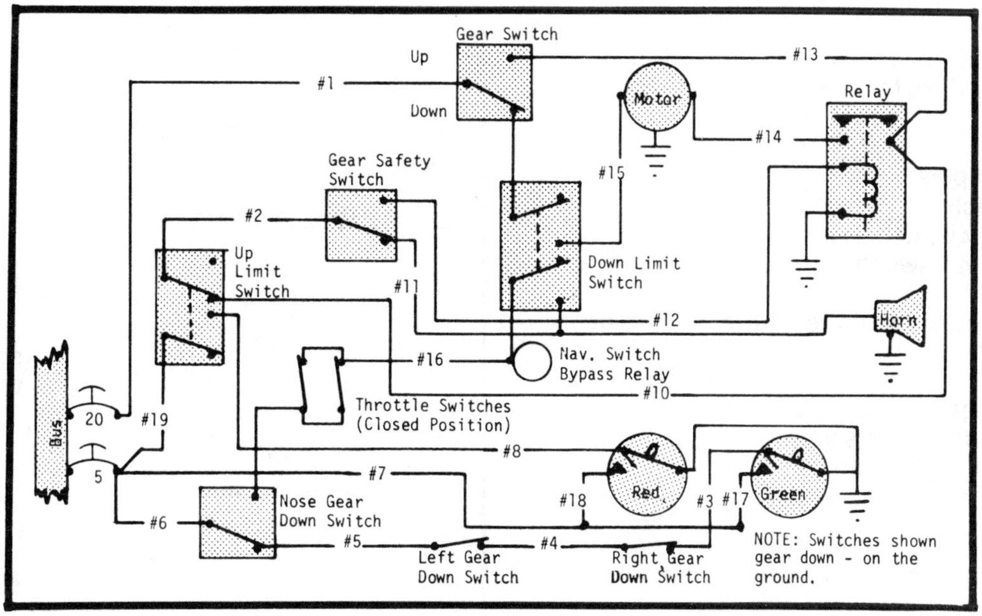

Figure 20

6966. Which of the following conditions is most likely to cause the landing gear warning signal to sound?

1— Landing gear not locked down and throttle advanced.
2— Landing gear locked down and throttle advanced.
3— Landing gear locked down and throttle retarded.
4— Landing gear not locked down and throttle retarded.

6967. (Refer to figure 21.) What will illuminate the amber indicator light?

1— Closing the nosewheel gear full retract switch.
2— Retarding one throttle and closing the left wheel gear locked down switch.
3— Closing the left or right wheel gear full retract switch.
4— Closing the nose, left and right wheel gear full retract switches.

6968. (Refer to figure 21.) When will the landing gear aural warning indicate an unsafe condition?

1— All gears up and one throttle retarded.
2— Any gear up and both throttles retarded.
3— Any gear up and one throttle retarded.
4— All gears down and locked and one or both throttles retarded.

6969. Where is the landing gear safety switch usually located?

1— On the main gear shock strut.
2— On the landing gear drag brace.
3— On the pilot's control pedestal.
4— On the landing gear selector.

6970. What safety device is actuated by the compression and extension of a landing gear strut?

1— Ground lockpins.
2— Uplock switch.
3— Downlock switch.
4— Ground safety switch.

6971. Which repair would require a landing gear retraction test?

1— Brake change.
2— Landing gear safety switch.
3— Red warning light bulb.
4— Gear downlock microswitch.

6972. Landing gear warning systems usually provide which of the following indications?

1— Red light for unsafe gear, no light for gear down, green light for gear up.
2— Green light for gear up and down, red light for unsafe gear.
3— A position indicator gauge plus green light for gear up, no light for gear down.
4— Red light for unsafe gear, green light for gear down, no light for gear up.

6973. In most modern hydraulically actuated landing gear systems, the order of gear and fairing door operation is controlled by

1— selector handles.
2— sequence valves.
3— limit switches.
4— microswitches.

6974. Which of the following landing gear warning devices are incorporated on retractable landing gear aircraft?

1— A visual indicator showing gear position.
2— A light which comes on when the gear is fully down and locked.
3— A light which comes on when the gear is fully up and locked.
4— A horn or other aural device and a red warning light.

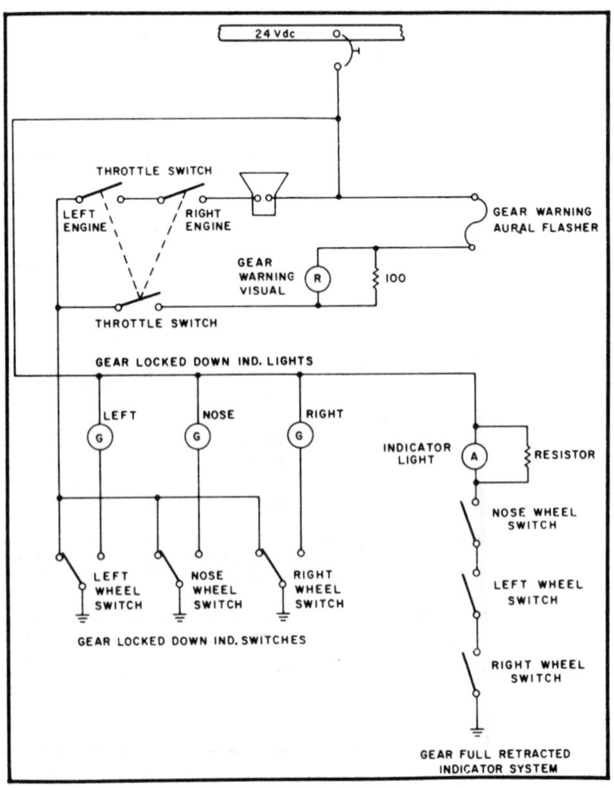

Figure 21

6975. When a landing gear safety switch on a main gear strut closes at liftoff, which system is deactivated?

1— Landing gear position system.
2— Antiskid system.
3— Pressurization system.
4— Aural warning system.

6976. The rotor in an autosyn remote indicating system uses

1— an electromagnet.
2— a permanent magnet.
3— an electromagnet and a permanent magnet.
4— neither an electromagnet nor a permanent magnet.

6977. The basic difference between an autosyn and a magnesyn indicating system is the

1— rotor.
2— transmitter.
3— receiver.
4— winding.

6978. The rotor in a magnesyn remote indicating system uses

1— a permanent magnet.
2— an electromagnet.
3— an electromagnet and a permanent magnet.
4— neither an electromagnet nor a permanent magnet.

6979. Microswitches are used primarily as limit switches to

1— limit generator output.
2— control electrical units automatically.
3— prevent overspeeding of a series motor.
4— prevent overcharging of a battery.

6980. Which of the following are some uses for a dc selsyn system?

A. Indicates position of retractable landing gear.
B. Indicates the angle of incidence of an aircraft.
C. Indicates the altitude of an aircraft.
D. Indicates cowl flaps or oil cooler door position.
E. Indicates fuel quantity.
F. Indicates the rate of climb of an aircraft.
G. Indicates position of wing flaps.

1— A, D, E, G.
2— B, C, D, E.
3— B, C, E, F.
4— A, B, G, F.

6981. (1) A dc selsyn system is a widely used electrical method of indicating a remote mechanical movement or position.

(2) A synchro-type indicating system is an electrical system used for transmitting information from one point to another.

Regarding the above statements,

1— only No. 1 is true.
2— only No. 2 is true.
3— both No. 1 and No. 2 are true.
4— neither No. 1 nor No. 2 is true.

6982. When installing pneumatic surface-bonded type deicer boots,

1— apply high-grade tire talc between the rubber and the wing skin.
2— remove all paint from the area to be covered by the deicer boot.
3— apply a solution of glycerin and water between the rubber and the wing skin.
4— apply a Silastic compound between the rubber and the wing skin.

6983. Which of the following are found in a laminated integral electrically heated windshield system?

A. Autotransformer.
B. Heat control relay.
C. Heat control toggle switch.
D. 24V dc power supply.
E. Indicating light.

1— A, B, D, E.
2— B, C, D, E.
3— A, B, C, E.
4— A, C, D, E.

6984. What is one check for proper operation of a pitot/static tube heater after replacement?

1— Ammeter reading.
2— Voltmeter reading.
3— Visual inspection of all connections.
4— Continuity check of system.

6985. What controls the inflation sequence in a pneumatic deicer boot system?

1— Boot construction.
2— Vacuum pump.
3— Distributor valve.
4— Suction relief valve.

6986. What is the source of pressure for inflating deicer boots on reciprocating engine aircraft?

1—Vane-type pump.
2—Gerotor pump.
3—Gear-type pump.
4—Piston-type pump.

6987. Which of the following regulates the vacuum of the air pump to hold the deicing boots deflated when the pneumatic deicing system is off?

1—Distributor valve.
2—Ejector.
3—Pressure regulator.
4—Suction relief valve.

6988. What may be used to clean deicer boots?

1—Unleaded gasoline.
2—Naphtha.
3—Neutral cleaning solvent.
4—Soap and water.

6989. Some aircraft are protected against airframe icing by heating the leading edges of the airfoils and intake ducts. When is this type of anti-ice system usually operated during flight?

1—Continuously while the aircraft is in flight.
2—In symmetric cycles during icing conditions to remove ice as it accumulates.
3—At all times while the outside air temperature is below freezing.
4—Whenever icing conditions are first encountered or expected to occur.

6990. Which of the following indications occur during a normal operational check of a pneumatic deicer system?

1—Relatively steady readings on the pressure and vacuum gauges.
2—Relatively steady readings on the pressure gauge and fluctuating readings on the vacuum gauge.
3—Fluctuating readings on the pressure gauge and relatively steady readings on the vacuum gauge.
4—Pressure and vacuum gauges will fluctuate as the deicer boots inflate and deflate.

6991. What method is usually employed to control the temperature of an anti-icing system using surface combustion heaters?

1—Thermo-cycling switches.
2—Flow control valves.
3—Thermostats in the cockpit.
4—Heater fuel shutoff valves.

6992. What is the purpose of the distributor valve in a deicing system utilizing deicer boots?

1—To regulate air pressure.
2—To equalize the air pressure to the left and right wings.
3—To sequence the deicer boot inflations.
4—To distribute anti-icing fluid to the deicer boots.

6993. What is the purpose of the oil separator in the pneumatic deicing system?

1—To protect the deicer boots from oil deterioration.
2—To remove oil from air exhausted from the deicer boots.
3—To prevent an accumulation of oil in the vacuum system.
4—To remove oil from the vacuum pump.

6994. Where are the heat sensors located on most aircraft with electrically heated windshields?

1—Imbedded in the glass.
2—Attached to the glass.
3—Around the glass.
4—Attached to the frame.

6995. Three possible sources of hot air for the operation of a wing thermal anti-icing system are

1—turbo-compressors, air storage tank, and vacuum pump.
2—engine bleed air, vacuum pump, and compressed air tank.
3—engine bleed air, combustion heaters, and augmentor tubes.
4—combustion heaters, augmentor tubes, and exhaust gases.

6996. What maintains normal windshield temperature control in an electrically heated windshield system?

1—Thermal overheat switches.
2—Autotransformers.
3—Thermistors.
4—Electronic amplifiers.

6997. Arcing in an electrically heated windshield panel usually indicates a breakdown in the

1—temperature-sensing elements.
2—autotransformers.
3—conductive coating.
4—thermal overheat switches.

6998. Which of the following directs vacuum to the deicer boots for holddown in flight?

1— Vacuum relief valve.
2— Ejector.
3— Distributor valve.
4— Pressure regulator and relief valve.

6999. How do deicer boots help remove ice accumulations?

1— By preventing the formation of ice.
2— By melting ice formations.
3— By breaking up ice formations.
4— By allowing only a thin layer of ice to build up.

7000. Why are the tubes in deicer boots alternately inflated?

1— Alternate inflation of deicer boot tubes keeps disturbance of the airflow to a minimum.
2— Alternate inflation of deicer boot tubes does not disturb airflow.
3— Alternate inflation of deicer boot tubes relieves the load on the air pump.
4— Alternate inflation of deicer boot tubes relieves the strain on the attach point.

7001. Carburetor icing may be eliminated by which of the following methods?

1— Alcohol spray and heated induction air.
2— Alcohol spray and heating induction duct.
3— Ethylene glycol spray and heated induction air.
4— Electrically heating air intake and ethylene glycol spray.

7002. Why should a chemical rain repellant not be used on a dry windshield?

1— It will etch the glass.
2— It will restrict visibility.
3— It will cause glass delamination.
4— It may crack the glass due to heat buildup.

7003. What is the principle of a windshield pneumatic rain removal system?

1— An air blast spreads a liquid rain repellant evenly over the windshield that prevents raindrops from clinging to the glass surface.
2— A liquid repellant is sprayed onto the windshield and uses the raindrops as a carrying agent to carry away the rain, keeping the glass surface clear.
3— An air blast forms a barrier that prevents raindrops from striking the windshield surface.
4— A pneumatic rain removal system is simply a mechanical windshield wiper system that is powered by pneumatic system pressure.

7004. What mixture may be used as a deicing fluid to remove frost from an aircraft surface?

1— Ethylene glycol and isopropyl alcohol.
2— Mild soap and water.
3— Methyl ethyl ketone and ethylene glycol.
4— Naphtha and isopropyl alcohol.

7005. What may be used to remove wet snow deposits from an aircraft?

1— A brush or a squeegee.
2— Hot air.
3— Warm water.
4— A chamois or a mop.

7006. What is used to prevent ice formation on a pitot tube?

1— An electric heating element built into the pitot head.
2— A ribbon heater installed around the pitot head.
3— A blanket-type heater installed on the pitot head.
4— A gasket heater installed at the base of the pitot head.

7007. What are three methods of anti-icing aircraft windshields?

A. Blanket-type heating system.
B. An electric heating element in the windshield.
C. Heated air circulating system.
D. Hot water system.
E. Windshield wipers and anti-icing fluid.
F. Ribbon-type heating system.

1— B, C, E.
2— A, B, F.
3— D, E, F.
4— B, C, D.

7008. What icing condition may occur in warm weather when there is no visible moisture present?

1— Glaze ice.
2— Rime ice.
3— Carburetor ice.
4— Wing leading edge ice.

7009. What should be used to melt the ice in a turbine engine if the compressor is immobile because of ice?

1— Deicing fluid.
2— Hot water.
3— Anti-icing fluid.
4— Hot air.

7010. What is used as a temperature-sensing element in an electrically heated windshield?

1— Resistor.
2— Thermistor.
3— Capacitor.
4— Condenser.

7011. In what area of an aircraft would you find a carbon monoxide detector?

1— Surface combustion heater compartment.
2— Cockpit and/or cabin.
3— Auxiliary power unit compartment.
4— Engine and/or nacelle.

7012. What occurs when a visual smoke detector is activated?

1— A warning bell within the indicator alarms automatically.
2— A lamp within the indicator illuminates automatically.
3— A lamp within the indicator extinguishes automatically.
4— The test lamp illuminates and an alarm is provided automatically.

7013. The three types of fire-extinguishing agents for aircraft interior fires are

1— water, methyl bromide, and carbon dioxide.
2— water, dry chemical, and methyl bromide.
3— water, carbon dioxide, and dry chemical.
4— water, dry chemical, and chlorobromomethane.

7014. When air samples contain carbon monoxide, portable carbon monoxide detectors containing yellow silica gel will turn which color?

1— Blue.
2— Green.
3— Pink.
4— Red.

7015. Smoke detection instruments are classified by their method of

1— construction.
2— installation.
3— maintenance.
4— detection.

7016. Smoke detectors which use a measurement of light transmissibility in the air are called

1— electromechanical devices.
2— photoelectrical devices.
3— visual devices.
4— electromeasuring devices.

7017. A contaminated carbon monoxide portable test unit would be returned to service by

1— heating the indicating element to 300 °F to reactivate the chemical.
2— installing a new indicating element.
3— evacuating the indicating element with CO_2.
4— taking no action. Unit is self-reactivating.

7018. Which fire-detection system measures temperature rise compared to a reference temperature?

1— Fenwal continuous loop.
2— Thermal switch.
3— Lindberg continuous element.
4— Thermocouple.

7019. A carbon dioxide (CO_2) hand-held fire extinguisher may be used on an electrical fire if

1— the siphon tube is rigid.
2— the horn is nonmetallic.
3— the handle is insulated.
4— the horn is nonmagnetic.

7020. The proper fire-extinguishing agent to use on an aircraft brake fire is

1— water.
2— carbon dioxide.
3— dry powder chemical.
4— carbon tetrachloride.

7021. Smoke in the cargo and/or baggage compartment of an aircraft is commonly detected by which instrument?

1— Visual scanner.
2— Chemical reactor.
3— Photoelectric cell.
4— Sniffer.

7022. Which statement is correct concerning the operation of a photoelectric smoke detector?

1— A photoelectric smoke detector measures the amount of smoke under a specific set of conditions.
2— A photoelectric smoke detector measures the amount of light available under a specific set of conditions.
3— A photoelectric smoke detector will warn only when smoke is present.
4— A photoelectric smoke detector is not affected by dust, soot, or other contaminants because it senses the difference between these and smoke.

7023. Why does the Fenwal fire-detection system use spot detectors wired parallel between two separate circuits?

1—A control unit is used to isolate the bad system in case of malfunction.
2—This installation is equal to two systems: a prime system and a reserve system.
3—A short may exist in either circuit without causing a false fire warning.
4—The dual terminal thermoswitch is used so that one terminal is wired to a bell, the other to a light.

7024. A fire-extinguisher container can be checked to determine its charge by

1—attaching a remote pressure gauge.
2—weighing the container and the remote control valve.
3—weighing the container and its contents.
4—a hydrostatic test.

7025. What is the color code for fire-extinguisher lines?

1—Brown.
2—Yellow.
3—Red and green.
4—Blue and yellow.

7026. The most common cause of false fire warnings in continuous-loop fire-detection systems is

1—improper routing or clamping of loops.
2—moisture.
3—dents, kinks, or crushed sensor sections.
4—contamination of sensors by aircraft fluids.

7027. A thermocouple in a fire-detection system causes the warning system to operate because

1—it generates a small current when heated.
2—heat decreases its electrical resistance.
3—it expands when heated and forms a ground for the warning system.
4—heat increases its electrical resistance.

7028. The thermocouple fire-warning system is activated by a

1—slowly overheated engine.
2—certain temperature.
3—core resistance drop.
4—rate-of-temperature rise.

7029. When used in fire-detection systems having a single indicator light, thermal switches are wired in

1—parallel with each other and in series with the light.
2—series with each other and the light.
3—parallel with each other and the light.
4—series with each other and parallel with the light.

7030. Built-in aircraft fire-extinguishing systems are ordinarily charged with

1—carbon monoxide and nitrogen.
2—freon and nitrogen.
3—carbon tetrachloride.
4—sodium bicarbonate.

7031. When the emergency shutoff valves are closed during an engine fire, the

1—fuel flow to the engine will be blocked.
2—fire-warning system will be deactivated.
3—fire extinguishers will automatically discharge.
4—fire-detection system will be deactivated.

7032. What maintenance is required on a fire-extinguisher system using a freon agent after it has been discharged?

1—Purge the system with nitrogen and replace the container.
2—Replace the container.
3—Purge the system with air and replace the container.
4—Flush the system with water and replace the container.

7033. What method is used to detect the thermal discharge of a built-in carbon dioxide fire-extinguisher system?

1—A discolorization of the yellow plastic disk in the thermal discharge line.
2—A rupture of the red plastic disk in the thermal discharge line.
3—The thermal plug missing from the side of the bottle.
4—A rupture of the green plastic disk in the thermal discharge line.

7034. The thermal switches of a bimetallic thermal-switch type fire-detection system are heat-sensitive units that complete circuits at a certain temperature. They are connected in

1— series with each other, and in series with the indicator lights.
2— parallel with each other, and in parallel with the indicator lights.
3— parallel with each other, but in series with the indicator lights.
4— series with each other, but in parallel with the indicator lights.

7035. (Refer to figure 22.) Using the chart, determine the temperature range for a fire-extinguishing agent storage container with a pressure of 330 PSIG. (Consider 330 PSIG for both minimum and maximum pressure.)

1— 47 to 73 °F.
2— 45 to 71 °F.
3— 47 to 71 °F.
4— 45 to 73 °F.

7036. (Refer to figure 22.) Determine what pressure is acceptable for a fire extinguisher when the surrounding area temperature is 33 °F.

1— 215 to 302 PSIG.
2— 214 to 301 PSIG.
3— 214 to 303 PSIG.
4— 215 to 301 PSIG.

CONTAINER PRESSURE VERSUS TEMPERATURE		
Temperature °F.	Container Pressure (PSIG)	
	Minimum	Maximum
-40	60	145
-30	83	165
-20	105	188
-10	125	210
0	145	230
10	167	252
20	188	275
30	209	295
40	230	317
50	255	342
60	284	370
70	319	405
80	356	443
90	395	483
100	438	523

Figure 22

7037. On a periodic check of fire-extinguisher containers, the pressure was not between minimum and maximum limits. What procedure should be followed?

1— Release pressure if above limits.
2— Replace the extinguisher container.
3— Increase pressure if below limits.
4— Leave it alone, as pressure will change with temperature change.

7038. In some fire-extinguishing systems, evidence that the system has been intentionally discharged is indicated by the absence of a

1— blue disk on the side of the fuselage.
2— red disk on the side of the fuselage.
3— green disk on the side of the fuselage.
4— yellow disk on the side of the fuselage.

7039. If a fire-extinguisher cartridge is removed from a discharge valve for any reason, it

1— can be used on another discharge valve assembly.
2— must be pressure checked.
3— is recommended that the cartridge be used only on the original discharge valve assembly.
4— cannot be used again.

7040. Which of the following are fire precautions which must be observed when working on an oxygen system?

A. Display "No Smoking" placards.
B. Provide adequate fire-fighting equipment.
C. Keep all tools and oxygen servicing equipment free from oil or grease.
D. Avoid checking aircraft radio or electrical systems.

1— A, C, and D.
2— A, B, and D.
3— B, C, and D.
4— A, B, C, and D.

7041. Which fire-extinguishing agent is considered to be the least toxic?

1— Carbon dioxide.
2— Bromotrifluoromethane (Halon 1301).
3— Dibromodifluoromethane (Halon 1202).
4— Bromochloromethane (Halon 1011).

7042. Maintenance of fire-detection systems includes the

1— repair of damaged sensing elements.
2— removal of excessive loop or element material.
3— recalibration of sensing elements.
4— replacement of damaged sensing elements.

7043. Which fire-extinguishing agents used in hand-operated extinguishers are acceptable for general use in aircraft?

1— Soda acid, aerosol foam, or carbon tetrachloride.
2— Dry chemical, water, or carbon tetrachloride.
3— Dry chemical, water, or carbon dioxide.
4— Carbon dioxide, aerosol foam, or soda acid.

7044. A squib, as used in a fire-protection system, is a

1— temperature-sensing device.
2— device for causing the fire-extinguishing agent to be released.
3— type of gauge for determining how much extinguishing agent remains in the tank.
4— probe used for installing frangible disks in extinguisher bottles.

APPENDIX 1

APPENDIX 1

ABBREVIATIONS AND REFERENCES

The abbreviations listed below are used to identify the reference associated with each test question.

AAV — Aircraft Air-conditioning (Vapor Cycle) EA-AAC-1 by AMP 1974
ABS — Aircraft Basic Science by Northrop Aeronautical Institute
AC — Advisory Circular
AEE — Aircraft Electricity and Electronics by Bent/McKinley
AES — Aircraft Electrical Systems by Jorch
AFM — Aircraft Fuel Metering Systems EA-FMS-1 by AMP 1975
AHS — Aircraft Hydraulic System EA-AH-1 by AMP 1975
AIS — Aircraft Instrument Systems EA-AIS by AMP 1976
AMD — Aircraft Maintenance by Daniel J. Brimm, Jr. and H. Edward Boggess
AMM — Aircraft and Missile Design and Maintenance Handbook by Charles A. Averbey
AMQ — Aero Mechanics Questionnaire
AMR — Aircraft Maintenance and Repair by Northrop Aeronautical Institute
AMR-5 — Aircraft Maintenance and Repair 5th Edition by Delp/Bent/McKinley
AOS — Aircraft Oxygen System EA-AOS-1 by AMP 1975
BI — Bonding Inspection—Quality Control Digest No. 5 FAA 1960
BS — Aircraft Bonded Structure EA-NMR by AMP 1977
CHE — Cherry Fasteners Literature by Townsend Textron
DAC — Douglas Aircraft Company Publication Composite Structure 1980
DAT — Dictionary of Aeronautical Terms by Dale Crane
EEA — Electricity and Electronics for Aerospace Vehicles by Northrop Institute of Technology
FAR — Federal Aviation Regulations
FPA — Fluid Power for Aircraft by Samuel W. Merrill
ITP — Integrated Training Program (General) ITP—G3 by AMP
ITP-A — Integrated Training Program (Airframe) EA-ITP-A by AMP 1982
ITP-B — Integrated Training Program (Airframe) EA-ITP-AB by IAP 1985
MMM — Manufacturer's Maintenance Manual
PAV — Powerplants for Aerospace Vehicles by Northrop Institute of Technology
SAM — Sabreliner Maintenance Manual
SMC — Aircraft Sheet Metal Construction and Repair EA-SMF by AMP 1978
TSO — Technical Standard Order

Item	*Reference*
6001.	AMR
6002.	AMR
6003.	AC 43.13-1A
6004.	AC 43.13-1A
6005.	AC 43.13-1A
6006.	AC 43.13-1A
6007.	AC 43.13-1A
6008.	AC 43.13-1A
6009.	AC 43.13-1A
6010.	AC 65-15A
6011.	AC 43.13-1A
6012.	AC 65-15A
6013.	AC 65-15A
6014.	AC 65-15A
6015.	AC 65-15A
6016.	AC 65-15A
6017.	AC 43.13-1A
6018.	AC 43.13-1A
6019.	AC 65-15A
6020.	AC 43.13-1A
6021.	AC 43.13-1A
6022.	AC 65-15A
6023.	AC 43.13-1A
6024.	AC 43.13-1A
6025.	AC 65-15A
6026.	AC 65-15A
6027.	AC 65-15A
6028.	AC 43.13-1A
6029.	AC 43.13-1A
6030.	AC 65-15A
6031.	AC 43.13-1A
6032.	AC 65-15A
6033.	AC 65-15A
6034.	AC 65-15A
6035.	FAR 45.29(c)
6036.	FAR 45.21
6037.	AC 65-15A
6038.	AC 65-15A
6039.	AC 65-15A
6040.	AC 65-15A
6041.	AC 65-15A
6042.	AC 65-15A
6043.	AC 43.13-1A
6044.	AC 43.13-1A
6045.	AC 65-15A
6046.	AC 65-15A
6047.	AC 65-15A
6048.	AC 65-15A
6049.	AC 65-15A
6050.	AC 65-15A
6051.	AC 65-9A
6052.	AC 65-15A
6053.	AC 65-15A
6054.	AC 65-9A
6055.	AC 65-9A
6056.	AC 65-9A
6057.	AC 65-9A
6058.	CEA
6059.	CHE
6060.	AC 65-15A
6061.	AC 65-15A
6062.	BS
6063.	BS
6064.	BS
6065.	BS
6066.	AC 43.13-1A
6067.	BS
6068.	BS
6069.	BS
6070.	AC 43.13-1A
6071.	AC 65-15A
6072.	AC 65-15A
6073.	AC 65-15A
6074.	AC 65-15A
6075.	AC 43.13-1A
6076.	AC 65-15A
6077.	AC 65-15A
6078.	AC 65-15A
6079.	AC 65-15A
6080.	AC 65-15A
6081.	AC 43.13-1A
6082.	BI
6083.	BS
6084.	BS
6085.	BS
6086.	BS
6087.	AC 43.13-1A
6088.	AC 43.13-1A
6089.	AC 43.13-1A
6090.	DAC
6091.	AC 65-15A
6092.	AC 43.13-1A

Item	Reference
6093.	TSO
6094.	AC 65-15A
6095.	AC 65-15A
6096.	AC 65-15A
6097.	FAR 23.853
6098.	AC 65-15A
6099.	AC 65-15A
6100.	AC 65-15A
6101.	AC 65-9A
6102.	AC 43.13-1A
6103.	AC 65-15A
6104.	AC 65-15A
6105.	AC 65-9A
6106.	AC 65-15A
6107.	AC 65-15A
6108.	AC 65-15A
6109.	AC 43.13-1A
6110.	AC 65-15A
6111.	AC 65-15A
6112.	AC 43.13-1A
6113.	AC 65-9A
6114.	AC 65-15A
6115.	AC 65-15A
6116.	AC 65-9A
6117.	AC 43.13-1A
6118.	AMR
6119.	AC 65-9A
6120.	AC 65-15A
6121.	AMR
6122.	AC 65-15A
6123.	AC 65-9A
6124.	AC 65-15A
6125.	AC 65-15A
6126.	AC 65-15A
6127.	AC 65-15A
6128.	AC 65-15A
6129.	AC 65-15A
6130.	AC 65-15A
6131.	SMC
6132.	AC 65-9A
6133.	AC 65-9A
6134.	AC 43.13-1A
6135.	AMM
6136.	AC 65-9A
6137.	AC 43.13-1A
6138.	AMM
6139.	AC 65-9A
6140.	AC 65-9A
6141.	AC 65-9A
6142.	AC 43.13-1A
6143.	AC 65-9A
6144.	AC 65-9A
6145.	AC 43.13-1A
6146.	AC 65-15A
6147.	AC 65-9A
6148.	AC 43.13-1A
6149.	AC 65-9A
6150.	AC 43.13-1A
6151.	AC 65-9A

Item	Reference
6152.	AC 65-15A
6153.	AC 65-15A
6154.	AC 65-15A
6155.	AC 65-15A
6156.	AC 65-9A
6157.	AC 65-9A
6158.	AC 65-15A
6159.	AC 65-15A
6160.	AC 65-15A
6161.	AC 65-15A
6162.	AC 65-15A
6163.	AC 65-15A
6164.	AC 65-15A
6165.	AC 65-15A
6166.	AC 65-15A
6167.	AC 65-15A
6168.	AC 65-15A
6169.	AC 65-15A
6170.	AC 65-15A
6171.	AC 65-15A
6172.	AC 65-9A
6173.	AC 65-15A
6174.	AC 65-15A
6175.	AC 65-15A
6176.	AC 65-15A
6177.	AC 65-15A
6178.	AC 65-15A
6179.	AC 65-15A
6180.	AC 65-15A
6181.	AC 65-9A
6182.	AC 65-15A
6183.	AC 65-15A
6184.	AC 65-15A
6185.	AC 65-15A
6186.	AC 65-15A
6187.	AC 65-15A
6188.	AC 43.13-1A
6189.	AC 43.13-1A
6190.	AC 65-15A
6191.	AC 65-15A
6192.	AC 65-15A
6193.	AMR-5
6194.	AC 65-15A
6195.	AC 65-15A
6196.	AC 43.13-1A
6197.	AC 65-15A
6198.	AC 65-15A
6199.	AC 65-15A
6200.	AC 43.13-1A
6201.	AC 65-15A
6202.	AC 65-15A
6203.	AC 65-15A
6204.	AC 65-15A
6205.	AC 65-15A
6206.	AC 65-15A
6207.	AC 65-15A
6208.	AC 65-15A
6209.	AC 65-15A
6210.	AC 65-15A

Item	Reference
6211.	AC 65-15A
6212.	AC 65-15A
6213.	AC 65-15A
6214.	AC 43.13-1A
6215.	AC 65-15A
6216.	AC 65-15A
6217.	AC 65-15A
6218.	AC 65-15A
6219.	AC 65-15A
6220.	AC 65-15A
6221.	AC 65-15A
6222.	AC 65-15A
6223.	AC 65-15A
6224.	AC 65-15A
6225.	AC 65-15A
6226.	AC 65-15A
6227.	AC 61-13B
6228.	AC 65-15A
6229.	AC 61-13B
6230.	AC 61-13B
6231.	AC 61-13B
6232.	AC 61-13B
6233.	AC 65-15A
6234.	AC 65-15A
6235.	AC 65-15A
6236.	AC 43.13-1A
6237.	AC 65-15A
6238.	AC 65-15A
6239.	AC 65-15A
6240.	AC 65-15A
6241.	AC 65-15A
6242.	AC 65-15A
6243.	AC 65-15A
6244.	AC 65-15A
6245.	AC 65-15A
6246.	AC 65-15A
6247.	AC 65-15A
6248.	AC 65-15A
6249.	AC 65-15A
6250.	AC 65-15A
6251.	AC 65-15A
6252.	AC 65-15A
6253.	AC 65-15A
6254.	AC 65-15A
6255.	AC 65-15A
6256.	AC 65-15A
6257.	AC 65-15A
6258.	AC 65-15A
6259.	AC 65-15A
6260.	AC 65-15A
6261.	AC 65-15A
6262.	AC 65-15A
6263.	AC 65-15A
6264.	AC 65-15A
6265.	AC 65-15A
6266.	AC 43.13-1A
6267.	AC 43.13-1A
6268.	AC 43.13-1A
6269.	AC 43.13-1A

Item	Reference
6270.	AC 43.13-1A
6271.	AC 65-9A
6272.	AC 43.13-1A
6273.	AC 65-15A
6274.	AC 65-15A
6275.	FAR 23.689(a)(1)
6276.	AC 65-15A
6277.	AC 43.13-1A
6278.	AC 65-15A
6279.	AC 65-15A
6280.	AC 43.13-1A
6281.	AC 43.13-1A
6282.	AC 43.13-1A
6283.	AC 65-15A
6284.	AC 65-9A
6285.	AC 65-15A
6286.	FAR 23.677(a)
6287.	AC 65-15A
6288.	AC 65-15A
6289.	AC 65-15A
6290.	AC 65-15A
6291.	AC 65-15A
6292.	AC 65-15A
6293.	AMD
6294.	AC 65-15A
6295.	AC 65-15A
6296.	AC 65-9A
6297.	AC 65-15A
6298.	AC 65-15A
6299.	AC 65-15A
6300.	AC 65-15A
6301.	AC 65-9A
6302.	AC 65-9A
6303.	AC 65-9A
6304.	AC 65-9A
6305.	FAR 43.7
6306.	FAR 43
6307.	FAR 91.169
6308.	FAR 43.11
6309.	FAR 91.169
6310.	FAR 43.7(b)
6311.	FAR 91.169
6312.	FAR 43.11
6313.	FAR 43
6314.	FAR 91.169
6315.	FAR 91.217
6316.	AMD
6317.	AC 65-15A
6318.	AC 65-15A
6319.	AC 65-15A
6320.	AC 65-15A
6321.	AC 65-15A
6322.	AC 65-15A
6323.	AC 43.13-1A
6324.	AC 65-15A
6325.	AC 65-15A
6326.	AC 65-15A
6327.	AMR
6328.	AC 65-15A

Item	Reference	Item	Reference	Item	Reference	Item	Reference
6329.	AC 65-15A	6388.	AMR	6447.	AHS	6506.	AC 65-15A
6330.	AC 65-15A	6389.	AC 65-15A	6448.	AHS	6507.	AC 65-15A
6331.	AC 65-15A	6390.	AC 65-9A	6449.	ITP-B	6508.	AMD
6332.	AC 65-15A	6391.	AHS	6450.	AC 65-15A	6509.	AC 65-15A
6333.	AC 65-15A	6392.	AC 65-15A	6451.	AC 65-15A	6510.	AC 65-15A
6334.	AC 65-15A	6393.	AC 65-15A	6452.	AC 65-15A	6511.	AC 65-15A
6335.	AC 65-15A	6394.	AHS	6453.	FPA	6512.	AC 65-15A
6336.	AC 65-15A	6395.	AHS	6454.	AC 65-15A	6513.	AC 65-15A
6337.	AC 65-15A	6396.	AHS	6455.	AC 65-15A	6514.	FPA
6338.	AC 65-15A	6397.	AHS	6456.	AC 65-15A	6515.	AHS
6339.	AC 65-15A	6398.	AMQ	6457.	AC 65-15A	6516.	AHS
6340.	AC 43.13-1A	6399.	AMQ	6458.	AC 65-15A	6517.	AC 43.13-1A
6341.	AC 65-15A	6400.	AC 65-15A	6459.	AC 65-15A	6518.	AC 65-15A
6342.	AC 65-15A	6401.	AC 65-9A	6460.	AMR	6519.	AC 65-15A
6343.	AC 65-15A	6402.	AC 65-9A	6461.	AC 65-15A	6520.	AC 65-15A
6344.	FAR 43	6403.	AC 43.13-1A	6462.	AC 65-15A	6521.	AC 65-15A
6345.	AC 65-15A	6404.	AC 65-9A	6463.	AC 65-15A	6522.	AC 65-15A
6346.	AC 65-15A	6405.	AC 65-15A	6464.	AC 65-15A	6523.	AC 65-15A
6347.	AC 65-15A	6406.	AC 65-15A	6465.	AC 65-9A	6524.	AC 65-15A
6348.	AC 65-15A	6407.	AC 65-15A	6466.	AC 65-9A	6525.	AC 65-15A
6349.	AC 65-15A	6408.	AC 65-15A	6467.	AC 65-9A	6526.	AC 65-15A
6350.	AC 65-15A	6409.	AC 65-15A	6468.	FPA	6527.	AC 65-15A
6351.	AC 65-15A	6410.	AC 65-15A	6469.	AC 65-15A	6528.	AC 65-15A
6352.	AC 65-15A	6411.	AC 65-15A	6470.	AC 65-15A	6529.	AC 65-15A
6353.	AC 65-15A	6412.	AC 65-15A	6471.	AC 65-15A	6530.	AAV
6354.	AC 65-15A	6413.	AC 65-15A	6472.	AC 65-15A	6531.	AAV
6355.	AC 43.13-1A	6414.	AC 65-15A	6473.	AC 65-15A	6532.	AAV
6356.	AC 65-15A	6415.	AC 65-15A	6474.	AMR	6533.	AC 65-15A
6357.	AC 65-15A	6416.	AC 65-15A	6475.	AC 65-15A	6534.	AAV
6358.	AC 65-15A	6417.	AC 65-15A	6476.	AC 65-15A	6535.	AC 43.13-1A
6359.	AMD	6418.	AC 65-15A	6477.	AC 65-15A	6536.	AC 43.13-1A
6360.	AC 65-15A	6419.	AC 65-15A	6478.	AC 65-15A	6537.	AC 43.13-1A
6361.	AC 65-15A	6420.	AC 65-9A	6479.	AC 65-15A	6538.	AOS
6362.	AC 65-15A	6421.	AC 65-15A	6480.	AC 65-15A	6539.	AC 65-15A
6363.	AC 65-15A	6422.	AC 65-9A	6481.	AC 65-15A	6540.	AC 65-15A
6364.	AC 65-15A	6423.	AC 65-9A	6482.	AC 65-15A	6541.	AC 65-15A
6365.	AC 65-15A	6424.	ITP-B	6483.	AMR	6542.	AC 65-15A
6366.	AC-65-15A	6425.	AHS	6484.	AC 65-15A	6543.	AC 65-15A
6367.	AC 43.13-1A	6426.	MMM	6485.	AC 65-15A	6544.	AC 65-15A
6368.	AC 65-15A	6427.	AC 65-9A	6486.	AC 65-15A	6545.	AC 65-15A
6369.	AC 65-15A	6428.	ITB-B	6487.	AC 65-15A	6546.	AC 65-15A
6370.	AC 65-15A	6429.	AC 65-15A	6488.	AC 65-15A	6547.	AC 65-15A
6371.	AC 65-15A	6430.	AC 65-15A	6489.	AC 65-15A	6548.	AC 65-15A
6372.	AC 65-15A	6431.	AC 65-15A	6490.	AC 65-15A	6549.	AC 65-15A
6373.	AC 65-15A	6432.	AC 65-15A	6491.	AC 65-9A	6550.	AC 65-15A
6374.	AC 65-15A	6433.	AC 65-15A	6492.	AC 65-15A	6551.	AC 65-15A
6375.	AC 65-15A	6434.	AC 65-15A	6493.	AC 65-15A	6552.	AC 65-15A
6376.	AC 65-15A	6435.	AC 65-15A	6494.	AC 65-15A	6553.	AC 65-15A
6377.	AC 65-15A	6436.	AC 65-15A	6495.	AMR	6554.	AC 65-15A
6378.	AC 65-15A	6437.	AC 65-15A	6496.	AC 65-15A	6555.	AC 65-15A
6379.	AC 65-15A	6438.	AC 65-15A	6497.	AC 65-15A	6556.	AC 65-15A
6380.	AC 65-15A	6439.	AC 65-15A	6498.	AC 65-15A	6557.	AC 65-15A
6381.	AC 65-15A	6440.	AC 65-15A	6499.	AC 65-15A	6558.	AC 65-15A
6382.	AC 65-15A	6441.	AC 65-15A	6500.	AC 65-15A	6559.	AC 65-15A
6383.	AC 65-15A	6442.	AC 65-15A	6501.	AC 65-9A	6560.	AC 65-15A
6384.	AC 65-15A	6443.	AC 65-15A	6502.	AC 65-15A	6561.	AMR-5
6385.	AC 65-15A	6444.	AC 65-15A	6503.	AC 65-15A	6562.	AC 65-15A
6386.	AMR	6445.	AC 65-15A	6504.	AC 65-15A	6563.	AC 65-15A
6387.	AC 65-15A	6446.	AHS	6505.	AC 65-15A	6564.	AC 65-15A

Item	Reference
6565.	AC 65-15A
6566.	AC 65-15A
6567.	AC 65-15A
6568.	AAV
6569.	AC 65-15A
6570.	AC 65-15A
6571.	AC 65-15A
6572.	AC 65-15A
6573.	AC 65-15A
6574.	AC 65-15A
6575.	AAV
6576.	AAV
6577.	AAV
6578.	AAV
6579.	AAV
6580.	AAV
6581.	AAV
6582.	AAV
6583.	AAV
6584.	AAV
6585.	AAV
6586.	AC 65-15A
6587.	ITP-B
6588.	ITP-B
6589.	ITP-B
6590.	AC 65-15A
6591.	AC 65-15A
6592.	AC 65-15A
6593.	AC 65-15A
6594.	AC 65-15A
6595.	AC 65-15A
6596.	AC 65-15A
6597.	AC 65-15A
6598.	AC 65-15A
6599.	AC 65-15A
6600.	AC 65-15A
6601.	AC 65-15A
6602.	AC 65-15A
6603.	AC 65-15A
6604.	AC 65-15A
6605.	AOS
6606.	AOS
6607.	AC 65-15A
6608.	AOS
6609.	ITP-B
6610.	ITP-B
6611.	AC 65-15A
6612.	AC 65-15A
6613.	AC 65-15A
6614.	AC 65-15A
6615.	AC 65-15A
6616.	FAR 23.1545
6617.	AC 65-15A
6618.	AC 65-15A
6619.	AC 65-15A
6620.	AC 65-15A
6621.	AC 65-15A
6622.	AC 65-15A
6623.	AC 65-15A

Item	Reference
6624.	AC 65-15A
6625.	AC 65-15A
6626.	AC 65-15A
6627.	FAR 65.81
6628.	AC 65-15A
6629.	FAR 65.81(a)
6630.	AC 65-15A
6631.	AC 65-15A
6632.	AIS
6633.	AC 65-15A
6634.	AC 65-15A
6635.	AC 65-15A
6636.	AC 65-15A and FAR 23
6637.	AC 65-15A
6638.	AC 65-15A
6639.	AC 65-15A
6640.	FAR 23.1327
6641.	AC 65-15A
6642.	AC 65-15A
6643.	FAR 23.1325
6644.	AC 65-15A
6645.	AC 65-9A
6646.	FAR 65.81
6647.	FAR 65.81
6648.	AC 65-15A
6649.	DAT
6650.	AC 65-15A
6651.	AC 65-15A
6652.	AC 65-15A
6653.	AC 65-15A
6654.	AC 65-15A
6655.	AC 65-15A
6656.	AC 65-15A
6657.	AC 65-15A
6658.	AC 65-15A
6659.	AC 65-15A
6660.	AC 65-15A
6661.	AC 65-15A
6662.	FAR 91.171
6663.	AC 65-15A
6664.	AC 65-15A
6665.	AC 65-15A
6666.	AC 65-15A
6667.	AC 65-15A
6668.	EEA
6669.	AC 65-15A
6670.	AC 65-15A
6671.	AC 65-15A
6672.	AC 65-15A
6673.	PAV
6674.	EEA
6675.	EEA
6676.	ITP-B
6677.	AC 65-15A
6678.	AC 65-15A
6679.	AC 43.13-2A
6680.	AC 65-15A
6681.	AC 65-15A

Item	Reference
6682.	AC 65-15A
6683.	AC 65-15A
6684.	AC 65-15A
6685.	AC 65-15A
6686.	AC 65-15A
6687.	AC 91-44
6688.	AC 65-15A
6689.	AC 65-15A
6690.	AC 65-15A
6691.	AC 65-15A
6692.	AC 65-15A
6693.	AC 43.13-2A
6694.	AC 65-15A
6695.	AC 65-9A
6696.	AC 65-15A
6697.	AC 43.13-2A
6698.	AC 65-15A
6699.	AC 43.13-2A
6700.	AC 65-15A
6701.	AC 43.13-2A
6702.	AC 43.13-2A
6703.	AC 65-15A
6704.	AC 65-15A
6705.	AC 65-15A
6706.	AC 65-15A
6707.	AC 65-15A
6708.	AC 65-15A
6709.	AC 65-15A
6710.	AC 65-15A
6711.	AC 43.13-2A
6712.	AC 65-15A
6713.	AC 65-9A
6714.	AC 65-9A
6715.	FAR 23.1001
6716.	FAR 23.1001
6717.	AC 65-9A
6718.	AC 65-9A
6719.	AC 65-9A
6720.	ITP-B
6721.	ITP-B
6722.	MMM
6723.	AC 65-9A
6724.	AC 65-9A
6725.	AC 65-9A
6726.	AC 65-9A
6727.	AC 65-9A
6728.	AC 43.13-1A
6729.	ITP-B
6730.	AC 65-9A
6731.	AC 65-9A
6732.	AC 65-9A
6733.	AC 65-9A
6734.	AC 65-9A
6735.	AC 65-9A
6736.	AC 65-9A
6737.	AC 65-9A
6738.	FAR 23.965(a)(1)
6739.	AC 65-9A
6740.	AC 65-9A

Item	Reference
6741.	AC 65-9A
6742.	AC 43.13-1A
6743.	AC 65-9A
6744.	AC 65-9A
6745.	AFM
6746.	AC 43.13-1A
6747.	AFM
6748.	AFM
6749.	AC 43.13-1A
6750.	AC 43.13-1A
6751.	AC 43.13-1A
6752.	AC 65-9A
6753.	AC 43.13-1A
6754.	AC 65-9A
6755.	AC 65-9A
6756.	AC 65-9A
6757.	AC 65-9A
6758.	AC 65-9A
6759.	AC 65-9A
6760.	AC 65-15A
6761.	AC 65-9A
6762.	AC 65-9A
6763.	AC 65-9A
6764.	AC 65-9A
6765.	AC 65-9A
6766.	AC 65-9A
6767.	AC 65-9A
6768.	AC 65-9A
6769.	AC 65-9A
6770.	AC 65-9A
6771.	AC 65-9A
6772.	AC 65-9A
6773.	FAR 23.1337
6774.	AC 65-9A
6775.	AC 65-9A
6776.	AC 65-9A
6777.	AC 65-9A
6778.	AC 65-9A
6779.	AC 65-9A
6780.	AC 65-9A
6781.	AC 65-9A
6782.	AC 65-9A
6783.	AC 65-9A
6784.	AC 65-9A
6785.	AC 65-9A
6786.	AC 65-12A
6787.	AC 65-12A
6788.	AC 65-12A
6789.	AC 65-9A
6790.	AC 65-9A
6791.	AC 65-9A
6792.	AC 65-9A
6793.	AC 65-9A
6794.	AC 43.13-2A
6795.	AC 65-9A
6796.	FAR 23.951(b)
6797.	AC 65-9A
6798.	FAR 25.1557
6799.	AC 65-9A

Item	Reference
6800.	AC 65-9A
6801.	AC 65-9A
6802.	AC 65-9A
6803.	AC 65-9A
6804.	FAR 23.1557
6805.	AC 65-9A
6806.	AC 65-9A
6807.	AC 65-9A
6808.	AC 65-9A
6809.	AC 43.13-1A
6810.	AC 65-9A
6811.	AC 65-9A
6812.	AC 65-9A
6813.	AC 65-9A
6814.	AC 65-9A
6815.	AC 65-9A
6816.	AC 65-9A
6817.	AC 65-9A
6818.	AC 43.13-1A
6819.	AC 65-9A
6820.	ITP-A
6821.	AC 65-9A
6822.	AC 65-9A
6823.	AES
6824.	AC 65-9A
6825.	AC 65-9A
6826.	AC 65-9A
6827.	AC 65-9A
6828.	AC 65-9A
6829.	AC 65-9A
6830.	AC 65-9A
6831.	AC 65-9A
6832.	AC 65-9A
6833.	AC 65-9A
6834.	AC 65-9A
6835.	AC 65-9A
6836.	AC 65-9A
6837.	AC 65-9A
6838.	AC 65-9A
6839.	AC 65-9A
6840.	AC 65-15A
6841.	AC 65-9A
6842.	AC 65-9A
6843.	AC 65-9A
6844.	EEA
6845.	AC 65-15A
6846.	FAR 23.1385
6847.	FAR 23
6848.	AC 65-9A
6849.	AC 65-9A
6850.	AC 65-15A
6851.	AC 65-9A
6852.	ITP-G3
6853.	ITP-G3
6854.	SAM
6855.	SAM
6856.	ITP-G3
6857.	ITP-G3
6858.	ITP-G3
6859.	ITP-G3
6860.	ITP-G3
6861.	ITP-G3
6862.	AC 65-9A
6863.	AC 65-9A
6864.	AC 65-9A
6865.	AC 65-15A
6866.	AC 65-15A
6867.	AC 43.13-1A
6868.	AC 43.13-1A
6869.	AC 65-9A
6870.	AC 65-15A
6871.	AC 65-15A
6872.	AC 65-15A
6873.	AC 65-9A
6874.	AC 65-9A
6875.	AC 65-15A
6876.	AC 65-9A
6877.	AC 43.13-1A
6878.	AC 43.13-1A
6879.	AC 43.13-1A
6880.	AC 65-15A
6881.	AC 65-15A
6882.	AC 65-9A
6883.	AC 43.13-1A
6884.	AC 43.13-1A
6885.	AC 43.13-1A
6886.	AC 43.13-1A
6887.	AC 65-15A
6888.	AC 65-15A
6889.	AC 65-9A
6890.	AC 65-15A
6891.	AC 43.13-1A
6892.	AC 43.13-1A
6893.	AC 43.13-1A
6894.	AC 43.13-1A
6895.	AC 65-9A
6896.	AC 65-15A
6897.	AC 65-9A
6898.	AC 43.13-1A
6899.	AC 65-15A
6900.	AC 65-9A
6901.	AC 43.13-1A
6902.	AC 43.13-1A
6903.	AC 65-15A
6904.	AC 43.13-1A
6905.	AC 65-15A
6906.	AC 43.13-1A
6907.	AC 65-9A
6908.	AC 43.13-1A
6909.	AC 65-15A
6910.	AC 65-9A
6911.	AC 65-15A
6912.	AC 65-9A
6913.	AC 65-9A
6914.	AC 65-9A
6915.	FAR 23.1385
6916.	EEA
6917.	AC 65-9A
6918.	AC 65-9A
6919.	EEA
6920.	AC 65-9A
6921.	AC 65-9A
6922.	AC 43.13-2A
6923.	AC 65-9A
6924.	AC 65-9A
6925.	AC 65-9A
6926.	AC 65-9A
6927.	AC 65-9A
6928.	AC 65-9A
6929.	AC 43.13-1A
6930.	AC 65-9A
6931.	AC 65-15A
6932.	AC 65-9A
6933.	AC 65-9A
6934.	AC 65-9A
6935.	AC 43.13-2A
6936.	AC 65-9A
6937.	AC 65-9A
6938.	AC 65-9A
6939.	AC 65-9A
6940.	AC 65-9A
6941.	AC 65-9A
6942.	AC 65-9A
6943.	AC 65-9A
6944.	AC 65-9A
6945.	AC 65-9A
6946.	AC 65-9A
6947.	AC 65-9A
6948.	AEE
6949.	AC 65-9A
6950.	AC 65-15A
6951.	AC 65-15A
6952.	AC 65-15A
6953.	AC 65-15A
6954.	AC 65-15A
6955.	AC 65 15A
6956.	FAR 23.1323
6957.	AC 65-15A
6958.	ITP-A
6959.	AC 65-15A
6960.	AC 65-15A
6961.	AC 65-15A
6962.	AC 65-15A
6963.	AC 65-15A
6964.	AC 65-15A
6965.	AC 65-15A
6966.	AC 65-15A
6967.	AC 65-15A
6968.	AC 65-15A
6969.	AC 65-15A
6970.	AC 43.13-1A
6971.	AC 65-15A
6972.	AC 65-15A
6973.	AC 65-15A
6974.	AC 65-15A
6975.	AC 65-15A
6976.	EA-AIS
6977.	EA-AIS
6978.	EA-AIS
6979.	AC 65-9A
6980.	AC 65-15A
6981.	AC 65-15A
6982.	AC 65-15A
6983.	AC 65-15A
6984.	AC 43.13-1A
6985.	AC 65-15A
6986.	AC 65-15A
6987.	AC 65-15A
6988.	AC 65-15A
6989.	AC 65-15A
6990.	AC 65-15A
6991.	AC 65-15A
6992.	AC 65-15A
6993.	AC 65-15A
6994.	AC 65-15A
6995.	AC 65-15A
6996.	AC 65-15A
6997.	AC 65-15A
6998.	AC 65-15A
6999.	AC 65-15A
7000.	AC 65-15A
7001.	AC 65-15A
7002.	AC 65-15A
7003.	AC 65-15A
7004.	AC 65-15A
7005.	AC 65-15A
7006.	AC 65-15A
7007.	AC 65-15A
7008.	AC 65-15A
7009.	AC 65-15A
7010.	AC 65-15A
7011.	AC 65-15A
7012.	AC 65-15A
7013.	AC 65-15A
7014.	AC 65-15A
7015.	AC 65-15A
7016.	AC 65-15A
7017.	AC 65-15A
7018.	AC 65-15A
7019.	AC 65-15A
7020.	AC 65-9A
7021.	AC 65-15A
7022.	AC 65-15A
7023.	AC 65-15A
7024.	AC 65-15A
7025.	AC 65-15A
7026.	AC 65-15A
7027.	AC 65-15A
7028.	AC 65-15A
7029.	AC 65-15A
7030.	AC 65-15A
7031.	AC 65-15A
7032.	AC 65-15A
7033.	AC 65-15A
7034.	AC 65-15A
7035.	AC 65-15A

Item	***Reference***
7036.	AC 65-15A
7037.	AC 65-15A
7038.	AC 65-15A
7039.	AC 65-15A
7040.	AC 65-15A
7041.	AC 65-15A
7042.	AC 65-15A
7043.	AC 65-15A
7044.	AC 65-15A